Michael Merker

Superconducting integrated THz receiver

Karlsruher Schriftenreihe zur Supraleitung Band 028

HERAUSGEBER

Prof. Dr.-Ing. M. Noe
Prof. Dr. rer. nat. M. Siegel

Eine Übersicht aller bisher in dieser Schriftenreihe erschienenen Bände finden Sie am Ende des Buches.

Superconducting integrated THz receiver

by
Michael Merker

KIT Scientific Publishing

Karlsruher Institut für Technologie
Institut für Mikro- und Nanoelektronische Systeme

Superconducting integrated THz receiver

Zur Erlangung des akademischen Grades eines Doktor-Ingenieurs von der KIT-Fakultät für Elektrotechnik und Informationstechnik des Karlsruher Instituts für Technologie (KIT) genehmigte Dissertation

von Dipl.-Ing. Michael Merker

Tag der mündlichen Prüfung: 24. Juni 2019
Hauptreferent: Prof. Dr. rer. nat. Michael Siegel
Korreferent: Prof. Dr.-Ing. Thomas Zwick

Impressum

KIT Scientific Publishing

Karlsruher Institut für Technologie (KIT)
KIT Scientific Publishing
Straße am Forum 2
D-76131 Karlsruhe

KIT Scientific Publishing is a registered trademark of Karlsruhe Institute of Technology.
Reprint using the book cover is not allowed.

www.ksp.kit.edu

Print on Demand 2019 – Gedruckt auf FSC-zertifiziertem Papier

ISSN 1869-1765
ISBN 978-3-7315-0970-7
DOI 10.5445/KSP/1000097713

Preface

The present work has been carried out at the Institute of Micro- and Nanoelectronic Systems (IMS) at the Karlsruhe Institute of Technology (KIT), Karlsruhe, Germany.

References

The citations within this work are subdivided into three different citation styles, indicating their respective origins.

- The cited literature is numbered serially by arabic numerals:

  [1].

- The publications that emanated during this thesis are indicated by the first letters of the author's last names and the year of publication:

  [MBV+17].

- This work has benefited from the contribution of students. These contributions are referred to with the first letters of the students' last name and the year of completion of the respective thesis:

  [Boh17].

The cited literature is appended, sorted in the order of appearance in the present work. Subsequently, an overview on the publications resulting from this work is given in chronological order. This is followed by the students' thesis in alphabetical order along with the international conferences, on which this work has been presented. The latter is presented in chronological order.

Karlsruhe, April 2019 Michael Merker

Danksagung

Diese Arbeit konnte in der vorliegenden Form durch die vielfältige Unterstützung von unterschiedlichen Seiten durchgeführt werden, an die ich an dieser Stelle meinen Dank richten möchte.

Zunächst gilt besonderer Dank Prof. Dr. M. Siegel, der diese Arbeit durch die Annahme der Promotion überhaupt ermöglicht hat. Darüber hinaus waren die vielen Gespräche sehr hilfreich. Das mir, von Beginn an und über die gesamte Promotionsdauer hinweg, permanent entgegengebrachte Vertrauen weiß ich sehr zu schätzen.

Prof. Dr. T. Zwick danke ich für die Übernahme des Korreferats.

Weiterhin seien die Mitarbeiter und Ex-Mitarbeiter des Instituts erwähnt. An erster Stelle Dr. S. Wünsch und Dr. K. Ilin, die mit ihrer Zeit und ihren immer offenen Ohren halfen, die verschiedensten Herausforderungen, die im Laufe der Zeit aufkamen, zu lösen. Kostya, durch deine große Geduld in Kombination mit deiner langjährigen Erfahrung in technologischen Fragestellungen konnten wir alle Herausforderungen meistern und das hohe technologische Niveau aufrecht erhalten, auch durch die "black streaks" hindurch. Vielen Dank. Stefan, ich danke dir für die Unterstützung in den verschiedensten Bereichen, von den Tücken der Verwaltungsbürokratie bis hin zu den fundamentalsten Fragen der Hochfrequenztechnik. Deine kritische, manchmal bewusst ketzerische Art zu fragen, hat mich oft zum Umdenken bewegt und mir somit ermöglicht, bestimmte Fragestellungen aus anderen Perspektiven zu betrachten. Dies in Kombination mit der angenehmen Zusammenarbeit hat mir die Arbeit deutlich erleichtert. K. Gutbrod möchte ich ebenfalls für die gute Zusammenarbeit danken, insbesondere bei der Durchführung des Handhold-Projekts oder der Neuaufsetzung des Praktikums, wenn wir mal wieder Feuerwehr sein mussten. Durch die ruhige und stets konstruktive Art konnten wir die "Brände immer löschen". Weitere technische Unterstützung habe ich auch von A. Stassen erfahren. Ob wieder eine Vorpumpe getauscht, der Mask-Aligner repariert oder eine chemische Fragestellung gelöst werden musste, konnte ich mich auf deine Unterstützung verlassen, vielen Dank Alex. Auch der IT-Support soll nicht unerwähnt bleiben. Frank, vielen Dank insbesondere für deine Mitarbeit im Handhold-Projekt, auch wenn dafür mal ein Wochenende dran glauben musste. Ein besonderer Dank gilt auch D. Duffner, mit deren Hilfe die bürokratischen Hürden immer locker genommen werden konnten. Vielen

Dank an alle weiteren Mitarbeiter des IMS für die gute, kollegiale Atmosphäre und die permanente Bereitschaft zur gegenseitigen Unterstützung. In alphabetischer Reihenfolge M. Arndt, I. Charaev, S. Dörner, D. Henrich, M. Hofherr, A. Kuzmin, J. Meckbach, J. Raasch, A. Scheuring, A. Schmid, E. Schmidt, A. Stockhausen, P. Thoma, P. Trojan. Besonders erwähnen möchte ich hierbei meinen JJ-Vorgänger: Max, du hast mich bestens ins IMS und in die JJ-Technologie eingeführt und somit auch einen großen Anteil an der erfolgreichen Durchführung dieser Arbeit. Nicht nur innerhalb des Instituts, auch außerhalb hatten wir immer eine gute Zeit. Vielen Dank! Ein weiterer Dank geht an Steffen. Ob bei fachlichen Diskussionen, der Organisation diverser "team-building-events" oder dem Entwurf und Erbauen des "Tesserakts", es war immer kurzweilig. Vielen Dank auch an Dr. K. Die Zusammenarbeit mit dir in den verschiedensten Projekten hat mir immer viel Spaß gemacht. Die vielen fachlichen Diskussionen waren immer hilfreich. Durch deine positive und konstruktive Art konnten wir neue Ideen schnell in die Tat umsetzen, wodurch die Idee der unterätzten Brücken zum fertigen Bolometer gebracht werden konnte, und das alles nebenbei. Alex, auch dir möchte ich für die gute Zusammenarbeit danken. Ob beim gemeinsamen Betreuen von studentischen Arbeiten oder den vielen fachlichen Gesprächen über die Hochfrequenztechnik. Darüber hinaus bin ich dir sehr dankbar für den Entwurf der Antennen der Mischer. Die Arbeit wäre in diesem umfassenden Ausmaß ohne die Mitwirkung der Studenten nicht möglich gewesen. Daher möchte ich mich auch bei euch für die gute Mithilfe bedanken. In chronologischer Reihenfolge: C. Karcher, C. Dischke, C. Bohn, S. Brenner, K. Fischer, T. Busch, M. Matura, M. Marholz, F. Nonnenmacher, M. Völlinger, L. Reiser und S. Schmid. Besonders hervorheben möchte ich hierbei Christian Bohn. Durch deine lanjährige Mitarbeit in der "Junction-group", angefangen bei deiner Bachelorarbeit über die Hiwi-Tätigkeit, die anschließende Masterarbeit und die darauffolgende Arbeit am IMS hast du einen maßgeblichen Anteil an meiner Arbeit. Danke dafür. Ich bin sicher, du hast am IHE den gleichen Spaß, auch ohne die Technologie und die Junctions, und wünsche dir viel Erfolg bei deiner Promotion.

Von Stellen außerhalb des IMS hat diese Arbeit ebenfalls profitiert. Daher möchte ich mich insbesondere bei der gesamten Tübingen-Gruppe um Prof. Dr. R. Kleiner und Prof. Dr. D. Kölle für die gute Zusammenarbeit bedanken. Im Besonderen Reinhold, Dieter, Edward, Matze, Rosina und Benni. Ich danke euch für die vielen guten Gespräche und die gute Zeit auf den zahlreichen Konferenzen, die es uns trotz der Distanz ermöglicht haben, diese enge Kollaboration aufrechtzuerhalten.

Auch mein privates Umfeld hat einen erheblichen Anteil am Gelingen dieser Arbeit. Allen voran meine Eltern. Auf eure permanente Unterstützung für meine Vorhaben, nicht nur während des Studiums und der Promotion, konnte ich mich immer verlassen. Eure wahnsinnige Geduld bei Erklärungen, wenn ich es mal wieder genauer wissen wollte, hat mit Sicherheit zu großen Teilen dazu beigetragen, dass ich diesen Weg eingeschlagen habe. Vielen Dank. Weiterhin meine beiden Brüder: Andi und Chris, auch bei euch weiß ich mit Sicherheit, dass ihr immer hinter mir steht und mich unterstützt. Für unser gutes Verhältnis bin ich euch unendlich dankbar. Darüber hinaus möchte ich mich auch bei meinen Freunden bedanken, die immer für die notwendige Ablenkung und Aufmunterung gesorgt haben: Ali, Carsten, Désirée, Dom, Isabel, Marc, Mosely, Nina und Tobi. Danke für die gute Zeit! Ein besonderer Dank geht an meine Freundin Nicole. Du hast es sicher als erstes erfahren, wenn die Technologie nicht so wollte, wie ich. Vielen Dank für deine Geduld und deine Unterstützung, insbesondere während der Anfertigung dieser Arbeit.

Kurzfassung

Seit der Entdeckung der Supraleitung durch Heike Kommerlingh Onnes im Jahre 1911 waren supraleitende Systeme Gegenstand intensivster Forschung. Insbesondere im Bereich der rauscharmen Signaldetektion kommen die Vorteile von gekühlten supraleitenden Komponenten in besonderem Maße zum Vorschein. Die weltweit empfindlichsten Empfänger werden daher auf Basis von supraleitenden Komponenten realisiert.

Die Anwendung von Supraleitern im Zusammenspiel mit dem im Jahre 1962 entdeckten Josephsoneffekt ermöglichen die Realisierung einer Vielzahl an Bauteilen. Die erfolgreichsten Bauteile und Anwendungen sind hierbei zum Beispiel die Spannungsstandardisierung mit einer bis dahin unerreichten Genauigkeit. Darüber hinaus können empfindlichste Magnetfeldsensoren auf Basis von sogenannten Superconducting Quantum Interference Devices (SQUIDs) realisiert werden. Die Entwicklung von Rapid Single Flux Quantum (RSFQ) basierter Logik ermöglichte darüber hinaus auch die Realisierung von logischen Operationen mit einer Taktfrequenz, welche die der heutigen Halbleitertechnologie um ein Vielfaches übertrifft. Dies und auch die Realisierung von Quantencomputern, basierend auf Quantumbits (QBits), sind Gegenstand der aktuellen Forschung.

Das drei-Schichtsystem, das den Josephsoneffekt ermöglicht, besteht aus zwei Supraleitern, welche durch eine dünne Barriere voneinander getrennt sind. Diese Barriere kann durch unterschiedlichste Materialien realisiert werden. Ohne Beschränkung der Allgemeinheit werden in dieser Arbeit jedoch lediglich Isolatoren als Barrierenmaterial betrachtet. Diese Supraleiter-Isolator-Supraleiter (SIS) Strukturen bilden die Basis für die Realisierung von extrem empfindlichen Heterodynempfängern. Für einen solchen Heterodynempfänger wird ein Bauteil mit möglichst nicht-linearer Kennlinie benötigt. Dieses nicht-lineare Bauteil wird mit einem lokal generierten Oszillatorsignal und dem zu empfangenden Signal beaufschlagt. Aufgrund der Nichtlinearität entstehen Intermodulationsprodukte der beiden Eingangssignalfrequenzen. Dieser Vorgang wird als Mischvorgang und das nicht-lineare Bauteil als Mischer bezeichnet. Da SIS-Kontakte hochgradig nicht-lineare Kennlinien aufweisen, eignen sich diese in besonderem Maße für einen Einsatz als Mischer. Insbesondere im Bereich des empfindlichen Empfangs von Signalen im sub-mm Wellenlängenbereich erweisen sich die SIS-Mischer als überlegen gegenüber jeder anderen Technologie.

Zusätzlich zu Mischer und Lokaloszillator wird auch eine Antenne benötigt, welche das zu empfangende Signal aus dem Freiraum in eine leitungsgebundene Übertragung überführt. Um eine effiziente Signalübertragung von Antenne zu Mischer zu erzielen, werden meist verlustarme, supraleitende Übertragungsleitungen verwendet. Nicht-integrierte Mischersysteme werden mit einem separaten, externen Lokaloszillator realisiert. Um einen supraleitenden, integrierten Empfänger zu realisieren, muss der Lokaloszillator mit dem Mischer und der Antenne auf einen gemeinsamen Chip integriert werden. Im Falle von SIS Mischern bietet sich die Realisierung des Lokaloszillators als Flux-Flow Oszillator an, da dieser in der identischen Technologie hergestellt wird. Aufgrund der gleichen Technologie kann die Herstellung zeitsparend durchgeführt werden und es kann auf eine Doppelprozessierung des gleichen Chips mit unterschiedlichen Technologien verzichtet werden. Das aktuell am häufigsten verwendete Supraleitermaterial ist reines Niob (Nb). Aufgrund der hohen Reproduzierbarkeit und der chemischen Stabilität wird damit das drei-Schichtsystem Niob/Aluminium-Aluminiumoxid/Niob (Nb/AlO_x/Nb) realisiert. In dieser Technologie wurden integrierte Empfänger bereits realisiert und erfolgreich eingesetzt.

Der Einsatz dieser Empfänger ist jedoch auf einen Frequenzbereich begrenzt welcher durch die Energielücke der verwendeten Supraleiter bestimmt ist. Daher können die auf Niob basierten integrierten Empfänger nur bis zu einer oberen Grenzfrequenz von ca. 700 GHz betrieben werden. Um die Grenzfrequenz zu erhöhen, müssen alternative supraleitende Materialien verwendet werden. Ein vielversprechendes supraleitendes Material ist Niobnitrid (NbN), welches eine theoretisch maximale Frequenz von 1.4 THz aufweist. Die Herstellung von NbN ist jedoch deutlich aufwändiger und schwieriger als bei Nb. Insbesondere das Substratmaterial spielt bei der Herstellung von qualitativ hochwertigen NbN-Schichten eine entscheidende Rolle. Das Substratmaterial, auf dem die mit Abstand hochwertigsten Schichten realisiert werden können, ist Magnesiumoxid (MgO). Bisher konnten auf anderen Substratmaterialien ohne Zwischenschicht keine vergleichbaren NbN-Filme abgeschieden werden. Aufgrund der hygroskopischen Eigenschaften von MgO wäre ein anderes Substratmaterial jedoch von Vorteil.

Daher soll im Rahmen der vorliegenden Arbeit die Abscheidetechnologie für qualitativ möglichst hochwertige NbN Dünnfilme auf Silizium- und Saphirsubstraten entwickelt und optimiert werden . Dies soll mit Hilfe eines drei-Kammer Sputtersystems erreicht werden. Das Hauptkriterium ist hierbei die erzielte Energielücke der resultierenden Filme. Darüber hinaus sollte die Oberflächenrauhigkeit der 200 nm dicken Schichten möglichst gering ausfallen, um eine elektrisch dichte Barriere zu ermöglichen. Nur wenn beide Kriterien erfüllt werden, können qualitativ hochwertige SIS-Kontakte realisiert werden. Da der komplexe Depositionsprozess eine Vielfalt an beeinflussenden Prozessparametern enthält, kann im Allgemeinen nicht nur einer der beiden Filmeigenschaften exklusiv beeinflusst werden, sondern nimmt in der Regel auch Einfluss auf das jeweils andere Qualitätskriterium. Daher wurde eine Optimierungsstrategie entwickelt, welche eine gezielte Beeinflussung der Filmeigenschaften durch bestimmte Prozessparameter erlaubt. Durch die Optimierung konnte die Energielücke der NbN Filme insbesondere auf Saphir deutlich verbessert werden, sodass diese Schichten einen Betrieb der Empfänger bei Frequenzen > 1 THz ermöglichen sollten.

Um ein möglichst hohes Maß an Flexibilität bei dem Entwurf der Hochfrequenzschaltungen zu erhalten, wurde der Strukturierungsprozess untersucht und optimiert. Die Optimierung zielt hierbei auf die zuverlässige Realisierung von möglichst geringen lateralen Abmessungen ab, ohne dabei die Prozessierungsdauer stark zu erhöhen. Es kann gezeigt werden, dass Strukturgrößen bis hin zu 500 nm bis in die obersten Schichtebenen realisiert werden können.

Auf Basis der ermittelten supraleitenden Eigenschaften der NbN Filme und der lateralen Strukturierungsgrenzen können Hochfrequenzsimulationen durchgeführt werden, um Schaltungen zu entwickeln, welche den Betrieb eines supraleitenden integrierten Empfängers ermöglichen. Dafür ist es essentiell, die Materialparameter bei hohen Frequenzen zu bestimmen und in die Simulationen zu integrieren. Eine weitere Herausforderung besteht in der korrekten Repräsentation der beiden aktiven Elemente in der Simulationssoftware. Mittels dreidimensionalen Feldsimulationen werden die Einzelteile der Hochfrequenzschaltungen entwickelt. Aufgrund der hohen Komplexität des Gesamtsystems eines vollintegrierten Heterodynempfängers werden mit den entworfenen Einzelteilen zunächst Teilsysteme realisiert, um bestimmte Fragestellungen an vereinfachten Prototypen untersuchen zu können. Die Fragestellungen beziehen sich hierbei sowohl auf die zuvor bestimmten Materialparameter, als auch auf die korrekte Implementierung des Mischers und des Flux-Flow Oszillators in die Simulation. Durch die gute Übereinstimmung der gemessenen mit den simulierten Eigenschaften dieser Prototypen, konnten die Simulationen verifiziert werden.

Auf dieser Basis wurde ein vollintegrierter Empfänger für eine Betriebsfrequenz von 750 GHz entworfen. Trotz der guten Übereinstimmung zwischen Simulation und Messung der Prototypen, zeigt dieser vollintegrierte Empfänger jedoch nicht das erwartete Verhalten. Es kann lediglich eine Leistungsübertragung von Flux-Flow Oszillator zu Mischer bei deutlich niedrigeren Frequenzen nachgewiesen werden. Trotz des unerwarteten Verhaltens wurden mehrere dieser Bauteile auf ihre Eigenschaften hinsichtlich der Strahlungseinkopplung über die Antenne hin untersucht. Diese optische Charakterisierung zeigt eine Signalübertragung bei Frequenzen bis zu 777 GHz. Eine Erklärung für dieses Verhalten konnte auch durch eine Analyse möglicher Ursachen bisher noch nicht gefunden werden. Der Frequenzbereich, in welchem eine Leistungsübertragung zwischen Flux-Flow Oszillator und Mischer nachweisbar ist, erstreckt sich von 300 bis 400 GHz. Daher wurden diese Empfänger in diesem Frequenzbereich betrieben, und der Empfängerbetrieb konnte durch externe Strahlung in Verbindung mit dem Flux-Flow Oszillatorsignal erfolgreich nachgewiesen werden.

In der vorliegenden Arbeit wurde somit ein Herstellungsprozess entwickelt und optimiert, um qualitativ hochwertige SIS-Kontakte auf Saphirsubstraten zu realisieren. Durch die Optimierung des Abscheideprozesses von NbN Dünnfilmen konnte die Filmqualität deutlich verbessert werden. Die damit realisierten Josephson Kontakte zeigen ebenso eine deutliche Verbesserung der Eigenschaften, insbesondere der Energielücke. Die erzielten Materialeigenschaften wurden in eine Hochfrequenzsimulation integriert, mittels derer eine Schaltung für einen integrierten Empfänger erstellt wurde. Diese Schaltungen wurden realisiert, und der Betrieb eines integrierten supraleitenden Heterodynempfängers in NbN Technologie konnte demonstriert werden.

Contents

1 Introduction

The developments and improvements in signal detection significantly enhance the capabilities of other research fields such as radio astronomy, which includes but is not limited to the investigation of the material composition of objects and galaxies in outer space. This can be done by analyzing the radiation emitted from these objects, which is in many cases in the sub-mm wavelength range. The very low intensity of these radiation signals in this high-frequency range and the awareness, that there is no amplifier with a noise temperature low enough to reasonably process such a signal, raises the demand of ultra-sensitive detectors and receivers for radio astronomy.

The devices used for the detection of such signals can be subdivided into two groups: The first group consists of the incoherent or direct detectors, in which the detection mechanism is based on the rectification of an incoming signal. The most prominent representative of these devices is the bolometer which is the most sensitive detector for THz radiation known to day [1]. The output signal generation is hereby based on the transformation of the incoming radiation signal to thermal energy, which leads to an increased temperature of the bolometer. The temperature induced resistance change creates the output signal of the bolometer. This group of detectors suffers from their low spectral resolution

$$R = \frac{\nu}{\Delta\nu}, \tag{1.1}$$

where the frequency of the signal of interest is given by ν and the bandwidth of the output signal $\Delta\nu$. For direct detectors, R is typically in the range of $R \approx 10$ [1]. Nevertheless, due to their high sensitivity, bolometer type direct detectors are widely used in systems which do not require a high spectral resolution rather than ultra-high sensitivity.

Since a significant effort of radio astronomy is dedicated to the recognition of single spectral lines of molecules [1–7], devices with ultra-high spectral resolution at frequencies in the THz range are of particular interest. This requirement can be granted by the second major group: the coherent or superheterodyne receivers. The operation principle of such a receiver is based on the down conversion of an incoming signal to a much lower frequency range, the baseband. This frequency conversion is achieved by using a mixer, which is a device with a strong non-linearity in its transmission characteristic. The non-linear device is subjected to the small signal of interest and another much stronger signal, generated by a local oscillator. Due to the much lower frequency of baseband, commercial electronics can be used for the detection of the signal. Therefore, a high spectral resolution is achieved, with actual values ranging from $R = 10^6$ [1] to $R \leq 10^8$ [8].

As opposed to the direct signal detection, a superheterodyne receiver conserves both, the amplitude as well as the phase of the signal and is thus subject to the Heißenberg uncertainty principle. This results in a fundamental quantum limit to the sensitivity, which can however not be achieved by applying classical mixer diodes or even super Schottky diode mixers [9, 10]. A large improvement in sensitivity can be achieved by the employment of bolometers as mixing elements. The application of bolometers as direct detectors has been discussed above. However, if the time-constants of a bolometer are small enough to follow an incident radiation signal, this device can be employed as a mixer. The first bolometers employed as mixers suffered from a strongly limited intermediate frequency bandwidth of less than 1 GHz [11]. The development of hot-electron bolometers (HEBs), relaxed this extreme drawback, increasing the intermediate frequency to values around 6 GHz [12, 13]. Research is ongoing to further increase the intermediate frequency bandwidth [14]. HEB based devices are not limited in operation frequency and have been shown to exhibit noise levels at the order of the quantum limit.

The only mixer device with a higher sensitivity is the Superconductor-Insulator-Superconductor (SIS)-based quasiparticle mixer. After the deployment of these devices as mixers and the quantum generalization of the mixer theory by Tucker [15, 16], it became clear, that these devices are actually able to reach the quantum limit [9, 10], while offering larger intermediate frequency bandwidths as compared to the HEB based receivers. The SIS-based receivers can be operated up to an upper frequency limit imposed by the gap frequency of the employed superconductors. Due to their superior properties, this type of receiver is employed in virtually all radio-astronomical research facilities observing the mm and sub-mm wavelength range.

The first SIS-based heterodyne receivers employed in radio astronomy were realized with a Pb based technology. These devices suffered from chemical instability and strong limitations in operation frequency. The discovery of the niobium-aluminum oxide-niobium (Nb-/Al-AlOx/Nb) technology was a major improvement. This technology is nowadays the most mature superconductor-insulator-superconductor technology and used in a wide variety of applications. It was a key technology for the realization of reliable and chemically stable devices such as Superconducting Quantum Interference Devices (SQUIDs), Rapid Single Flux Quantum circuits (RSFQ) or voltage standards and also induced a significant improvement in the receiver technology, enabling the realization of ultra-low noise heterodyne receivers with significantly higher frequencies up to the gap frequency of niobium of 700 GHz. In order to overcome the limitations and further increase the operation frequency, research is nowadays focused on the substitution of niobium as electrode and transmission line material by other superconductors with higher gap frequencies such as niobium nitride (NbN) or niobium titanium nitride (NbTiN) [17–22]. The fabrication technology of these superconductors is much more challenging, which is why numerous hybrid attempts have been investigated and demonstrated. This includes the realization of the mixing element in the well-established niobium based technology, embedded in an RF network which is realized by superconductors with higher energy gaps. Other approaches also substitute one of the electrodes with a higher energy gap superconductor. The highest frequency will be achieved with the employment of superconductors with highest energy gaps for each metalization layer. Therefore, research is continuing, aiming at an increase in operation frequencies of the receivers by introducing superconductors with higher energy gaps. Another important criterion for the SIS mixers is their RF bandwidth, which is determined by the mixer capacity. In order to increase the mixer bandwidth, SIS trilayers with high critical current densities are of interest. However the aluminum oxide barrier in the Nb/AlO_x/Nb technology seems to exhibit a limitation in critical current density at around $15\,kA/cm^2$. In order to increase the critical current density, the aluminum oxide is replaced with aluminum nitride (AlN), which enabled a significant increase in the critical current density [23].

Apart from increasing the operation frequency and the bandwidth of a single pixel receiver, the overall size, weight and power consumption of such a receiver is another aspect that has to be considered. Especially air or space borne applications would significantly benefit from a system with low weight and low power consumption. Therefore, an integration of as many parts as possible on a single chip is beneficial. The integration of the passive components such as the antenna or the matching transformer poses no significant challenge due to the high quality of fabrication of planar structures and transmission line technology. However, the integration of a local oscillator requires the implementation of an active device, which generates THz radiation on the same substrate. This local oscillator can be realized by a flux-flow oscillator (FFO) which can be fabricated in the same technology as the SIS mixer. A system integrating a SIS-based mixer with a flux-flow oscillator is called a superconducting integrated receiver (SIR). The employment of a flux-flow oscillator as a local oscillator also offers the possibility to create an array of single pixels on a single chip, further decreasing the space consumption of the heterodyne instruments.

In the context of the possible improvements presented above, this work is dedicated to the development of an integrated receiver based on niobium nitride electrodes and an aluminum nitride barrier. The basics necessary for understanding the optimization processes are presented in chapter 2 and 3, including both, the high-frequency fundamentals as well as the physical background. The properties of the superconductors relevant for the optimization are introduced along with the optimization itself is presented in chapter 4. The optimization hereby focuses on the achievement of a maximized energy gap and a minimized surface roughness of the niobium nitride films. The complex deposition process is therefore investigated regarding various deposition parameters. The optimization strategy is presented along with the extracted dependencies. Furthermore, the aluminum nitride barrier deposition is discussed. Using these optimized deposition parameters, the resulting junctions exhibit significantly higher quality parameters as compared to those prior to the optimization. For the development of an integrated receiver based on the developed technology, the receiver is divided into single components, which can be optimized in simulations individually. The development of these single components is discussed in chapter 5. Due to the high complexity of an entire integrated receiver, initially, subsystems thereof are designed, fabricated and characterized, which is presented in chapter 6. With this gained knowledge, the characterization of a subsequently realized fully integrated receiver is done. This characterization is presented in chapter 7.

2 Fundamentals of the heterodyne principle and microwaves

The signal recognition by means of a heterodyne receiver is nowadays one of the most common receiving techniques employed in a wide variety of applications and measurement tools. This includes simple applications such as radio, but also more complex applications such as mobile phone communication, radar applications and high performance RF measurement tools such as network or spectrum analyzers or satellite communication. Due to the high level of frequency resolution, these receivers are even employed in radio astronomy and atmospheric research. This chapter shows the state-of-the-art millimeter and sub-millimeter wave detection after which the concept of heterodyne receivers is introduced, along with the necessary fundamentals of microwave technology and the most important quantities.

2.1 State-of-the-art of sub-mm wave detection

The unmatched performance of superheterodyne receivers resulted in an extensive investigation and development for their deployment in various research facilities covering the entire spectrum from a few tens of GHz up to exceeding 1 THz [19, 24–36]. Today, these receivers are present in almost all of the worlds millimeter and sub-millimeter wave observatories such as the ground based **A**tacama **L**arge **M**illimeter/Submillimeter **A**rray (ALMA), the airborne systems **TE**rahertz and sub-millimeter wave **li**mb **s**ounder (TELIS) and **S**tratospheric **O**bservatory **F**or **I**nfrared **A**stronomy (SOFIA) or the spaceborne Herschel Space Observatory. The state-of-the-art superheterodyne receiver technology can be demonstrated using the ALMA or the SOFIA instrumentation, where the latter incorporates the GREAT instrument (**G**erman **RE**ceiver for **A**stronomy at **T**erahertz Frequencies) which employs HEB based receivers to cover a frequency span from 1.25 THz to 5 THz with a spectral resolution as high as $R \leq 10^8$. The ALMA instrumentation uses SIS-based heterodyne instruments to cover a frequency range from 84 GHz up to 950 GHz with nearly quantum-limited sensitivity [9, 37].

2.2 Heterodyne detection

The fundamental principle of superheterodyne detection is based on the simultaneous transmission of two signals to a mixing element. This mixing element exhibits a strongly non-linear current-voltage characteristic. An illustration of such a setup is given in figure 2.1a. The two signals are the signal of interest (RF) and the local oscillator (LO) signal which are combined in a signal combiner and then simultaneously fed to the mixing element. The operation point of the mixer is set by the amount of incident power of the local oscillator. This process is called pumping.

Due to the non-linear current-voltage characteristic, the output signal of the mixer does not only contain the superimposed signals, but also the intermodulation products of the input signals. This includes among others, the difference in frequencies which is called the intermediate frequency (IF). If the local oscillator frequency was selected close to the frequency of the signal of interest, the intermediate frequency is in the range of a few GHz. This frequency can be extracted by filtering with low-pass or band-bass filters and can then be amplified and further processed using commercially available low-noise amplifiers and electronics. A receiver system with an external local oscillator, typically employs a Mylar beam splitter or diplexer as signal combiner. The mixing process can be described mathematically by analyzing the current in the mixer. Applying the local oscillator and the RF signal to the mixer results in a voltage change which can be expressed by

$$\Delta V = V_{\mathrm{LO}} \cos(\omega t) + V_{\mathrm{RF}} \cos((\omega + \Delta\omega)t), \tag{2.1}$$

in which $\omega = 2\pi\nu$ represents the frequency of the local oscillator. $\Delta\omega$ is the frequency difference between the local oscillator an the incoming signal. The signal amplitudes are given by V_{i}, where we assume $V_{\mathrm{RF}} << V_{\mathrm{LO}}$. Derivation of the Taylor series for $I(V_0 + \Delta V)$ with the bias voltage V_0 results in

$$\begin{aligned} I =& I(V_0) + \frac{\partial I}{\partial V} \cdot (V_{\mathrm{LO}} \cos(\omega t) + V_{\mathrm{RF}} \cos((\omega + \Delta\omega)t)) \\ &+ \frac{1}{2} \frac{\partial^2 I}{\partial V^2} \left(\frac{1}{2} V_{\mathrm{LO}}^2 + \frac{1}{2} V_{\mathrm{LO}}^2 \cos(2\omega t) + V_{\mathrm{LO}} V_{\mathrm{RF}} (\cos((2\omega + \Delta\omega)t)) + \cos(\Delta\omega t) \right). \end{aligned} \tag{2.2}$$

In (2.2), terms with higher orders are neglected. The first term represents the bias (DC) current, whereas the last term contains a part depending on the frequency difference of ν_{LO} and ν_{RF}. Figure 2.1b illustrates the down mixing process in the frequency domain. The red arrow represents the local oscillator signal, the blue spectra represent two sidebands which

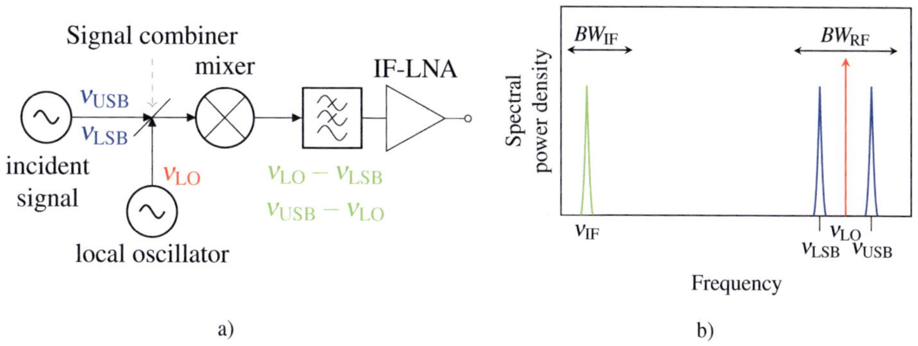

Figure 2.1: Schematic illustration of the heterodyne principle. (a) depicts the schematic setup comprising the incident signal fed to a mixing element along with a local oscillator signal. The intermediate frequency among others contains a signal with the difference frequency of the two input signals After a low pass filter, the signal is amplified and can then be processed. (b) depicts the operation of a heterodyne receiver in the frequency domain. The upper and lower side band signals are both down converted to the intermediate frequency.

are both down converted to the intermediate frequency band. This behavior can be seen directly from equation 2.2. These two bands are called upper sideband (USB) for $\nu_{RF} > \nu_{LO}$ and lower sideband (LSB) respectively. Since both, the USB as well as the LSB are down converted to the same IF frequency they are indistinguishable after down conversion. If one of the bands is to be suppressed, it has do be realized in the RF path by e.g. a sideband separating receiver. This is however not part of this work and the interested reader is referred to [1, 38].

The frequency range that can be covered by the receiver is called the RF bandwidth and determines the receiver operation frequency limits. In practical observatory facilities the frequency spectrum to be monitored is typically large and a single receiver can not cover this entire spectrum due to the limited bandwidth. Therefore, the large target spectrum is usually subdivided into different bands, where each band can be covered by a single receiver.

Another important property of the receiver is the bandwidth of the intermediate frequency (IF-bandwidth). This bandwidth has an immediate influence on the timing efficiency of the investigation of a certain spectrum [1]. If a spectrum of interest exceeds the intermediate frequency bandwidth, the local-oscillator frequency has to be adjusted accordingly multiple times in order to characterize the entire spectrum. Therefore, higher IF-bandwidths lead to shorter scan times to obtain a certain spectrum.

In order to characterize the noise properties of a receiver system, the amplifier chain has to be considered. It is common in amplifier theory to characterize an amplifier by the means of a noise temperature T_n. This value refers to the temperature of a fictive black body connected to the input of an ideal, noiseless amplifier producing the same amount of output signal as the noise power of the real amplifier. The noise power P_n produced by such a black body at the amplifier input is given by

$$\frac{P_n}{B} = k_B T_n, \tag{2.3}$$

where B is the bandwidth of the amplifier and k_B is the Boltzmann constant. Modeling the noise source as an input of the amplifier, the output power of the amplifier is then given by

$$P_{out} = G_{amp}\,(P_{in} + P_n) \tag{2.4}$$

with the amplifier gain G_{amp}, the signal input power P_{in} and the noise power P_n. The gain relates the power at the amplifier output to the input power and for a heterodyne receiver, it can be written as

$$G_{amp} = \frac{P_{RF}}{P_{IF}}. \tag{2.5}$$

The noise temperature can be expressed as [1]

$$T_n = \frac{T_{out}}{G_{amp}} - T_{in}, \tag{2.6}$$

with T_i representing the particular temperatures. An ideal amplifier with $P_n = 0$ leads to $T_{in} = T_{out}$, which can not be achieved by a heterodyne system, since it is subject to the Heißenberg uncertainty principle due to the phase conservation during down conversion. Thus, the minimum noise added to the input of the receiver is at least half a quantum of noise $(h\nu/2k_B)$ [1].

Unlike an amplifier, which per definition has a gain greater than unity, heterodyne receivers typically exhibit a gain smaller than unity due to intrinsic losses of the conversion process, such as the creation of harmonics. This is referred to by the term conversion loss. However, it has been predicted by Tucker [15, 39] that SIS-based heterodyne receivers can actually exhibit a conversion gain under certain circumstances, which has been proven experimentally in e.g. [24, 40, 41]. This result can only be obtained with an extremely sharp non-linearity which can be exhibited by SIS tunnel junctions. This result emphasizes the superiority of the SIS junctions in heterodyne receivers.

Conclusion

In this section, the fundamental concept of heterodyne receiving has been introduced. The mixing of a high frequency signal with another signal with a similar frequency at a device with a strongly non-linear current-voltage characteristic allows a signal detection with an extremely high spectral resolution. The down converted signal is at the order of a few GHz and can be further processed with commercial electronics. Due to the phase conservation of the incoming signal during down conversion, the sensitivity of a heterodyne receiver is limited to the quantum limit given by the Heißenberg uncertainty principle, which can not be reached by classical mixing elements such as diodes. The usage of superconducting elements such as HEBs or SIS mixers allows to closely approach this limit. However, a receiver is not only determined by its mixer properties, furthermore, the mixer is embedded into a microwave environment, which requires a low loss power transmission to the mixer in order to maximally exploit the mixer properties.

2.3 RF Power transmission and impedance matching

This chapter is dedicated to the power transmission of radio frequencies. The fundamental idea of impedance matching is introduced, as well as the impedance transformation using transmission lines.

2.3.1 Transmission lines

Transferring RF power between two points in space is achieved by the usage of transmission lines which can be cables or planar structures, that are specially designed for a high power transmission. In most cases, they are formed by two wircs, a conductor and a ground wire. Modeling such a transmission line in order to derive the particular properties is done using an equivalent circuit model as depicted in figure 2.2. In this equivalent circuit, an infinitely short part of a transmission line with two wires is considered. The most important properties are described as lumped elements, where the conductor wire is described by a resistor accounting for Ohmic losses due to the finite resistance of the material and an inductance which accounts for the inductance of the conductor wire. The two wires are typically separated from each other by a dielectric medium which is characterized by its permittivity ε_r and conductance G. This conductance accounts for the losses that occur in the dielectric. The capacitor represents the capacity determined by the geometrical setup of the transmission line as well as the dielectric properties.

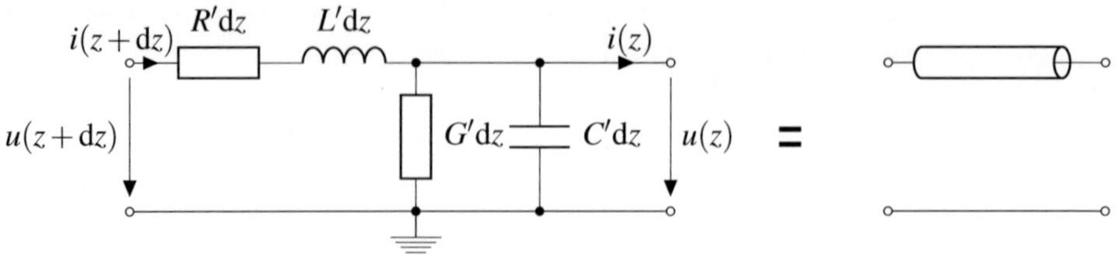

Figure 2.2: General equivalent circuit of a transmission line. A transmission line is modeled using the lumped elements given in the figure.

For the resulting circuit, the voltage and the current on a transmission line can be described by

$$\frac{\partial V(z)}{\partial z} = -(R' + j\omega L')I(z) \tag{2.7}$$

$$\frac{\partial I(z)}{\partial z} = -(G' + j\omega C')V(z). \tag{2.8}$$

These fundamental equations in transmission line theory are known as the telegrapher's equations. Differentiation of both equations and subsequent substitution yields wave equations for the voltage $V(z)$ and the current $I(z)$:

$$\frac{\partial^2 V(z)}{\partial z^2} - \gamma^2 V(z) = 0 \tag{2.9}$$

$$\frac{\partial^2 I(z)}{\partial z^2} - \gamma^2 I(z) = 0, \tag{2.10}$$

with $\gamma = \alpha + j\beta = \sqrt{(R + j\omega L)(G + j\omega C)}$ being the complex propagation constant of the wave in the transmission line. Solutions for (2.9) and (2.10) are given by

$$V(z) = V_0^+ \mathrm{e}^{-\gamma z} + V_0^- \mathrm{e}^{\gamma z} \tag{2.11}$$

$$I(z) = I_0^+ \mathrm{e}^{-\gamma z} + I_0^- \mathrm{e}^{\gamma z}, \tag{2.12}$$

using the convention with V^+ being the wave in positive z direction, whereas V^- describes the reflected wave and thus propagates in negative z direction. Applying (2.7) to (2.12), the ratio between voltage and current can be written as

$$\frac{V(z)}{I(z)} = \frac{R' + \mathrm{j}\omega L'}{\gamma} = \sqrt{\frac{R' + \mathrm{j}\omega L'}{G' + j\omega C'}} = Z_0. \tag{2.13}$$

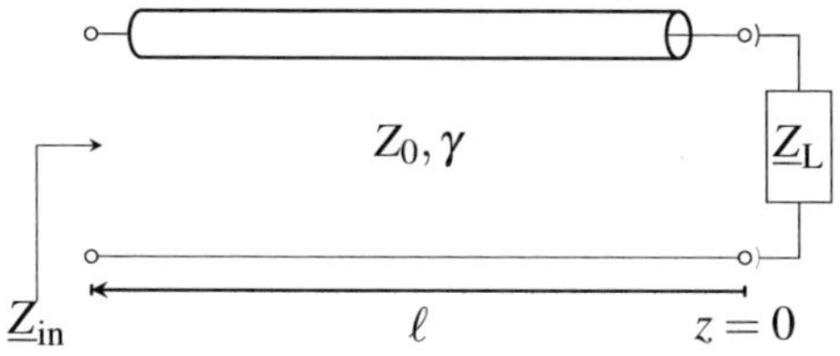

Figure 2.3: A transmission line connected to a load impedance $\underline{Z}_L$. The transmission line is characterized by its characteristic impedance Z_0, the propagation constant γ and its length ℓ.

This ratio has the physical unit of Ω and is therefore named the characteristic impedance Z_0. The voltage waveform in (2.11) corresponds to a time domain signal given by

$$v(z,t) = \left|V_0^+\right| \cos\left(\omega t - \beta z + \Phi^+\right) \mathrm{e}^{-\alpha z} \tag{2.14}$$

$$+ \left|V_0^-\right| \cos\left(\omega t - \beta z + \Phi^-\right) \mathrm{e}^{\alpha z} \tag{2.15}$$

with $\Phi^\pm$ being the phase angle of the complex voltage V_0. This equation represents an attenuated sinusoidal wave. The decay is described by the attenuation constant α and is given by $\mathrm{Re}(a + \mathrm{j}b)$

$$\alpha = \mathrm{Re}(\gamma) = \mathrm{Re}\left(\sqrt{(R' + j\omega L')(G' + j\omega C')}\right), \tag{2.16}$$

and the wavelength λ and phase velocity v_p are

$$\lambda = \frac{2\pi}{\beta} = \frac{2\pi}{\mathrm{Im}\left(\sqrt{(R' + \mathrm{j}\omega L')(G' + \mathrm{j}\omega C')}\right)} \tag{2.17}$$

$$v_\mathrm{p} = \lambda \nu \tag{2.18}$$

with the frequency ν. The relation between the reflected and the incident wave determines the voltage reflection coefficient Γ. This quantity can be used in order to qualify the amount of reflected power on a transmission line terminated by a certain load as depicted in figure 2.3

$$\Gamma = \frac{V_0^-}{V_0^+} = \frac{Z_1 - Z_2}{Z_1 + Z_2}, \tag{2.19}$$

where V_0^+ and V_0^- represent the voltage amplitudes of the incident and reflected waves respectively. Γ can be a complex value with an absolute value between -1 and 1. If the load impedance matches the characteristic impedance, this result in $\Gamma = 0$. In case of $\Gamma = \pm 1$, all power is reflected. Especially for multi-port devices, not only the reflection is an important quantity, but also the transmission from a specific port to another.

In order to quantify the transmission, it is convenient to use the scatter parameter ($\underline{S}$- parameter) matrix. The complex $\underline{S}$-parameter matrix of an N-port network is defined according to:

$$\begin{bmatrix} V_1^- \\ V_1^- \\ \vdots \\ V_N^- \end{bmatrix} = \begin{bmatrix} \underline{S}_{11} & \underline{S}_{12} & \cdots & \underline{S}_{1N} \\ \underline{S}_{21} & \underline{S}_{12} & \cdots & \underline{S}_{2N} \\ \vdots & & & \vdots \\ \underline{S}_{N1} & \underline{S}_{N2} & \cdots & \underline{S}_{NN} \end{bmatrix} \begin{bmatrix} V_1^+ \\ V_1^+ \\ \vdots \\ V_N^+ \end{bmatrix} \tag{2.20}$$

$$\left[V^-\right] = [\underline{S}] \left[V^+\right] \tag{2.21}$$

where the particular $\underline{S}$-parameter valued are calculated as

$$\underline{S}_{ij} = \left.\frac{V_i^-}{V_j^+}\right|_{V_k^+ = 0 \text{ for } k \neq j} . \tag{2.22}$$

The values $\underline{S}_{ij}$ are thus determined by driving a port i with an incident wave of voltage V_i^+ where the input at every other port is set to zero. All other ports should thus be terminated by matched loads. By monitoring the reflected wave V_i^-, the $\underline{S}$-parameter can be calculated and is typically given in decibels (dB). For a two port device, $\underline{S}_{11}$ equals the reflection coefficient Γ for this port. The power transmitted from port 1 to port 2 is given by $\underline{S}_{21}$.

Lossless transmission line

The calculations of the properties of transmission lines were done for the general case so far. This includes the consideration of the losses in the different materials. However in some cases, the lossless approach is sufficient for an accurate modeling of a transmission line. In this special case the losses in both, the conductor as well as the dielectric are neglected ($R' = G' = 0$). This approach is especially useful for low frequencies. In this case, the propagation constant becomes a purely imaginary number

$$\gamma = j\omega\sqrt{L'C'} . \tag{2.23}$$

This implies that $\alpha = 0$ and therefore results in a wave along the transmission line without attenuation. The characteristic impedance simplifies to

$$Z_0 = \sqrt{\frac{L'}{C'}} \tag{2.24}$$

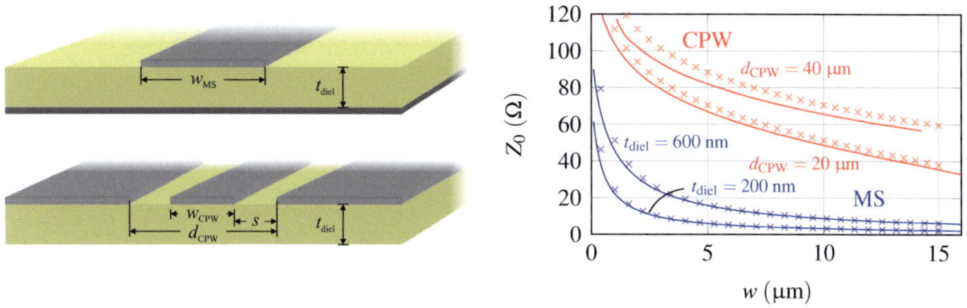

a) b)

Figure 2.4: Schematic cross section of microstrip transmission line (MS) and a coplanar waveguide (CPW) with characteristic impedances. (a) Schematic cross sections of a microstrip transmission line and a coplanar waveguide. (b) Example simulation of the achievable characteristic impedances of these two types of transmission line for varying conductor width w_i. The simulations with perfect electric conductors as metallization are depicted by the solid lines whereas the calculated impedances by the approximations given in (2.29) and (2.34) are depicted as crosses.

and the wavelength and phase velocity are given by

$$\lambda = \frac{2\pi}{\omega\sqrt{L'C'}} \tag{2.25}$$

$$v_p = \frac{\omega}{\beta} = \frac{1}{\sqrt{L'C'}}. \tag{2.26}$$

The microstrip and coplanar transmission lines

As seen in (2.13) and (2.24), the characteristic impedance of a transmission line depends on the inductance and the capacitance of the structure, which in turn are determined by the material properties and the geometry of the transmission line. In microelectronics, almost exclusively, the microstrip and coplanar transmission lines are used, since they can be miniaturized and integrated in planar technologies. Figure 2.4a shows the schematic cross sections of these two types of transmission lines which will be introduced in the following section.

The microstrip line is a transmission line type, where the ground plane is covered by a dielectric on which the flat conductor wire, the microstrip, is located. Due to the fact, that the dielectric does not fill the entire space (especially above the microstrip), a pure TEM wave such as in waveguides or coaxial cables can not propagate due to the different phase velocities $v_{\mathrm{p,air}}$ and $v_{\mathrm{p,diel}}$ in the different media

$$v_{\mathrm{p,air}} = c \tag{2.27}$$
$$v_{\mathrm{p,diel}} = \frac{c}{\sqrt{\varepsilon_{\mathrm{r}}}}, \tag{2.28}$$

but propagate as a quasi-TEM wave. This also affects the characteristic impedance, for which good approximations were found [38]:

$$Z_{0,\mathrm{MS\ eff}} = \begin{cases} \frac{60}{\sqrt{\varepsilon_{\mathrm{eff,MS}}}} \ln\left(\frac{8t_{\mathrm{diel}}}{w_{\mathrm{MS}}} + \frac{w_{\mathrm{MS}}}{4t_{\mathrm{diel}}}\right) & \text{for } w_{\mathrm{MS}}/t_{\mathrm{diel}} \leq 1 \\ \frac{120\pi}{\sqrt{\varepsilon_{\mathrm{eff,MS}}\left(\frac{w_{\mathrm{MS}}}{t_{\mathrm{diel}}} + 1.393 + 0.667\ln\left(\frac{w_{\mathrm{MS}}}{t_{\mathrm{diel}}} + 1.444\right)\right)}} & \text{for } w_{\mathrm{MS}}/t_{\mathrm{diel}} \geq 1 \end{cases} \tag{2.29}$$

with $\varepsilon_{\mathrm{eff,MS}}$ as an effective dielectric constant that can be estimated by

$$\varepsilon_{\mathrm{eff,MS}} = \frac{\varepsilon_{\mathrm{r}}+1}{2} + \frac{\varepsilon_{\mathrm{r}}-1}{2} \frac{1}{\sqrt{1+12\frac{t_{\mathrm{diel}}}{w_{\mathrm{MS}}}}}. \tag{2.30}$$

Using this estimation, also an effective wavelength and phase velocity can be calculated

$$\lambda = \frac{\lambda_0}{\sqrt{\varepsilon_{\mathrm{eff,MS}}}} \tag{2.31}$$
$$v_{\mathrm{p}} = \frac{c}{\sqrt{\varepsilon_{\mathrm{eff,MS}}}}, \tag{2.32}$$

where λ_0 equals the wavelength without dielectric medium between the conductors. Note that for a constant frequency, according to (2.31) and (2.30), the wavelength changes with different characteristic impedances due to the change in $\varepsilon_{\mathrm{eff}}$. The cross section of a coplanar waveguide is depicted as well in figure 2.4a. It is a fully planar technology and thus widely used. It consists of a center conductor, which is surrounded by a ground plane in a certain distance s. The resulting gap along with the width of the center conductor defines the characteristic impedance of this type of transmission line for given material parameters. Similar to the microstrip line, the coplanar wave guide only supports quasi-TEM waves. As for the microstrip line, an effective dielectric constant $\varepsilon_{\mathrm{eff}}$ is introduced [42] which is in this case

$$\varepsilon_{\mathrm{eff,CPW}} = \frac{\varepsilon_{\mathrm{r}}+1}{2}. \tag{2.33}$$

Table 2.1: Parameters for the calculation and simulation of the characteristic impedances.

	ν (GHz)	t_{diel} (nm)	d_{CPW} (μm)
MS:	600	200..600	-
CPW:	600	50000	20..40

With this value, the characteristic impedance can be estimated using

$$Z_{0,\text{CPW eff}} = \frac{30\pi}{\varepsilon_{\text{eff,CPW}}} \cdot \frac{K(k')}{K(k)}, \tag{2.34}$$

where $K(k)$ and $K(k')$ are the complete elliptic integral of the first kind with the modulus $k = \frac{w}{2s+w}$ and $k' = \sqrt{(1-k^2)}$ respectively. The wavelength and phase velocity are calculated in analogy to the microstrip case, however, the effective dielectric constant for the coplanar waveguide has to be used:

$$\lambda = \frac{\lambda_0}{\sqrt{\varepsilon_{\text{eff,CPW}}}} = \frac{\lambda_0}{\frac{\varepsilon_r+1}{2}} \tag{2.35}$$

$$v_p = \frac{c}{\sqrt{\varepsilon_{\text{eff,CPW}}}} = \frac{c}{\frac{\varepsilon_r+1}{2}}. \tag{2.36}$$

Figure 2.4b depicts the achievable impedance range for both microstrip and coplanar waveguides for typical geometries used in this work, showing the decrease in impedance for wider conductors. The graph contains both, values from the equations (2.29) and (2.34) as well as results obtained from 3D field simulations, where the latter are represented by the solid lines. Apparently, the calculated values fit the simulated values for $w_i > 2$ μm with only minor deviations.

This figure was created for values of t_{diel} and d_{CPW} given in table 2.1 which significantly differ from typical values found in literature for transmission lines geometries [WGMS15]or [43]. Microstrip transmission lines are typically realized using the entire substrate as dielectric, which is significantly thicker ($\approx 0.1 - 0.5$ mm) than the realizable dielectric thicknesses in the context of this work. This limited dielectric thickness stems from the geometrical setup given by the fabrication process and will be discussed in chapter 4. The graph in figure 2.4b furthermore contains the characteristic impedances of coplanar waveguides for given total widths ($d_{\text{CPW}} = w_{\text{CPW}} + 2 \cdot s$). This limited degree of freedom does not apply in general, but is selected in order to demonstrate the significantly higher characteristic impedances that can be achieved by using coplanar waveguides.

Conclusion

In this section, the transmission line is introduced along with the derivation of its general properties. The most important characteristic quantities are shown such as the reflection coefficient and the scattering parameters as a practical quantity to characterize RF systems. In the frame of this work, only planar transmission line types are considered. The two most important and most frequently used transmission line types are introduced, the microstrip transmission line and the coplanar waveguide, along with their particular properties. The characteristic quantities of transmission lines are introduced and discussed using lossless transmission lines. Furthermore, for these two transmission line types, approximations are given which allow the estimation of the characteristic impedances of transmission lines by means of the introduction of an effective permittivity.

2.3.2 Concept of impedance matching

In a microwave environment, one of the most important challenge is the power transmission from an RF generator to a load. This subject is schematically depicted in figure 2.5. The RF generator in the left part of the figure is assumed as an ideal generator with a series impedance $Z_g = R_g + jX_g$, with R being the resistance and X the susceptance of the complex impedance. It can be shown [38], that the maximum amount of power is transmitted from the generator to the load, if the condition of conjugate matching is met. In this case, the input impedance Z_{in} in figure 2.5 needs to obey

$$\underline{Z}_g = \underline{Z}_{in}^*, \tag{2.37}$$

where $\underline{Z}_{in}^*$ denotes the conjugate complex value of $\underline{Z}_{in}$. In this state, the maximum power is transferred from the generator to the load and is given by

$$P = \frac{1}{2}\left|V_g\right|^2 \frac{1}{4R_g}. \tag{2.38}$$

If $\underline{Z}_{in}$ does not obey the criterion given in equation (2.37), the impedance matching network needs to be designed in order to fulfill this condition. This is realized by a matching network which transforms the impedance seen by the generator as depicted in figure 2.5. This matching network needs to designed to the particular needs of the RF circuit. In the simplest case for purely real impedances, the matching network can simply contain a series or parallel resistor, so the sum of the resistances match the generator impedance. However, for complex load impedances, the matching network also needs to compensate for the complex part.

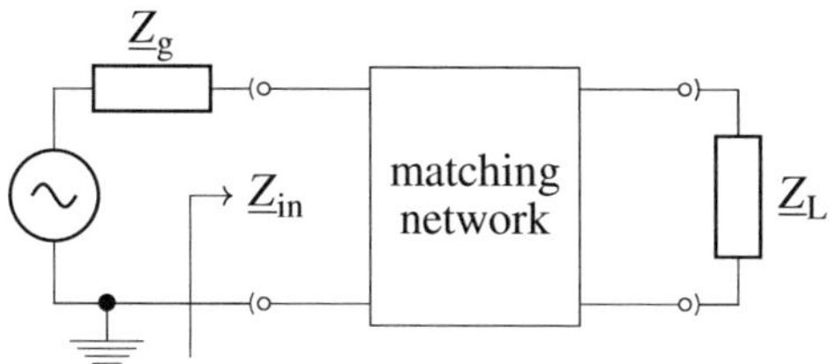

Figure 2.5: Equivalent circuit depicting a generator connected to a load impedance. The load impedance needs to be the conjugate complex of the generator output impedance for maximized power transmission. In general, this is not guaranteed, and a matching network is required.

This can be done by adding reactive elements such as a capacitance or inductance to the matching network with the impedances

$$\underline{Z}_{\text{cap}} = -\frac{j}{\omega C} \tag{2.39}$$
$$\underline{Z}_{\text{ind}} = j\omega L. \tag{2.40}$$

Obviously, the impedances of these elements are frequency dependent and therefore need to be calculated for the specific target frequency. This also implies that a perfect compensation of the load susceptance is usually only valid for one or more discrete frequencies rather than a continuous spectrum. As an example, a load with a capacitive impedance can be matched by a parallel inductor resulting in purely real impedance for the frequency

$$\nu_{\text{res}} = \frac{1}{2\pi}\sqrt{\frac{1}{LC}}. \tag{2.41}$$

There are usually multiple options to achieve the matched case for given generator and load impedances with typically different bandwidths. A detailed derivation of achieving matching with lumped elements can be found in [38].

2.3.3 Impedance transformation by transmission lines

In the previous section, transmission lines have been introduced with their properties, however not including the interaction of the transmission line with its environment. Now, a lossless transmission line connected to a load impedance is considered, as depicted previously in figure 2.3. The transmission line properties are defined by the characteristic impedance Z_0, the propagation constant β and the length ℓ. An arbitrary load impedance $\underline{Z}_L$ is connected at $z = 0$.

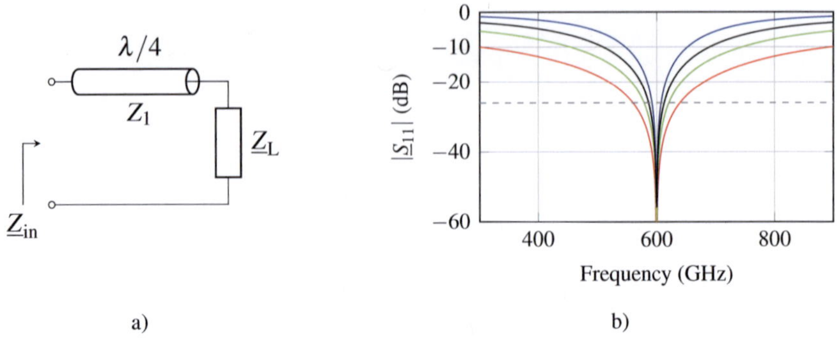

Figure 2.6: Illustration of the $\lambda/4$ transformation. (a) The equivalent circuit of the transmission line connected to the load impedance. The calculated reflection for this type of impedance transformation is depicted in (b) for different generator impedances at the input of the $\lambda/4$ transformer, ranging from 5 Ω to 50 Ω. The load impedance is 2 Ω.

In order to calculate the impedance seen at the input of the transmission line $\underline{Z}_{in}$, the ratio of the voltage to current at the point $z = -\ell$ needs to be evaluated. It can be shown [38] that the input impedance of the transmission line is given by

$$\underline{Z}_{in} = Z_0 \cdot \frac{\underline{Z}_L + jZ_0 \tan(\beta\ell)}{Z_0 + j\underline{Z}_L \tan(\beta\ell)}. \tag{2.42}$$

In the trivial case of $\underline{Z}_L = Z_0$, this equation simply yields $\underline{Z}_{in} = \underline{Z}_L$. This means, the input impedance seen by the source is independent on the length of the transmission line. Another setup, that does not alter the input impedance, is given with $\ell = \lambda/2 \Rightarrow \beta\ell = \pi$. In this case, the complex parts vanish and again, the input impedance equals the load impedance. However, in the general case $Z_0 \neq Z_L$, the load impedance is transformed by the transmission line depending on its length and its characteristic impedance. An important case is the quarter-wavelength transformer, which will be presented in the following section.

$\lambda/4$-transformer

The special case of a quarter-wavelength transformer is depicted schematically in figure 2.6a. This widely used case simply consists of a transmission line with a length $\ell = \lambda/4 \Rightarrow \beta\ell = \pi/2$ and a characteristic impedance Z_1, and is thus named a quarter-wavelength transformer. In this case, equation 2.42 yields

$$\lim_{\ell \to \lambda/4} \underline{Z}_{in}(\ell) = \frac{Z_1^2}{\underline{Z}_L}. \tag{2.43}$$

Table 2.2: Fractional bandwidth for different Z_g and a constant $Z_0 = 50\,\Omega$.

Z_g (Ω)	Z_l (Ω)	$\frac{\Delta\nu}{\nu_0}$ (%)
5	3.16	13.5
10	4.47	7.1
20	6.32	4.5
50	10	2.7

Assuming a real load impedance, $\underline{Z}_{in}$ is purely real with a value that is defined by the load impedance and the characteristic impedance of the transmission line.

Single-section transformer The $\lambda/4$ transmission line can thus be used as an impedance transformer for e.g. impedance matching. In this context, using equation (2.43), the characteristic impedance of the transmission line needs to be selected for a given load impedance and a desired input impedance according to

$$Z_l = \sqrt{\underline{Z}_{in}\underline{Z}_L}. \tag{2.44}$$

Due to the simplicity of this impedance transformation, it is widely used. The drawback of this method is that the impedance matching is given perfectly only for one single frequency and its harmonics. The matching strongly deteriorates for an increasing frequency difference between the real operation frequency and the design frequency, resulting in a strongly limited bandwidth. Figure 2.6 dcpicts $|\underline{S}_{11}|$ for the matching of a 2 Ω load to different input impedances from 5 to 50 Ω. The fractional bandwidth can be calculated according to [44] with

$$\frac{\Delta\nu}{\nu_0} = 2 - \frac{4}{\pi}\cos^{-1}\left|\frac{2|\underline{S}_{11,max}|\sqrt{Z_L Z_{in}}}{(Z_L - Z_{in})\sqrt{1-|\underline{S}_{11,max}|^2}}\right| \tag{2.45}$$

with the maximum accepted reflection $|\underline{S}_{11,max}|$. An example for the fractional bandwidths of the quarter-wavelength transformation with a selected $|\underline{S}_{11,max}| = -26$ dB is given in table 2.2, assuming a generator with an impedance $\underline{Z}_g$ connected to the input.

Multisection transformer If the bandwidth of a $\lambda/4$ single stage transformer is not sufficient for a particular application, this can be overcome by multistage transformers. These transformers consist of n $\lambda/4$-transformer where each of them has a particular characteristic impedance, as depicted in figure 2.7a.

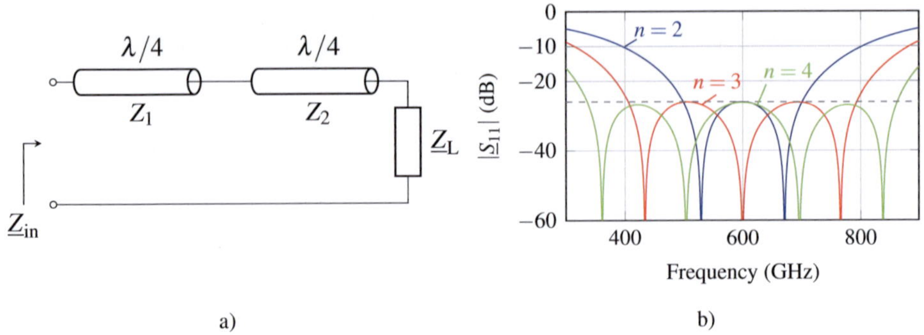

Figure 2.7: Equivalent circuit and bandwidth of a Chebychev matching transformer. (a) The equivalent circuit of a two-stage Chebychev type matching transformer, comprising two single $\lambda/4$ transformations. (b) The calculated reflection of multi-stage Chebychev type matching transformers for $n = 2, 3, 4$.

An incident wave is reflected partly at each transition between different characteristic impedances. By selecting the particular characteristic impedances of the different stages, this effect can be exploited to achieve different transmission characteristics such as a binomial or a Chebychev characteristic.

Compared to the binomial transformer, the Chebychev transformer exhibits a larger bandwidth, which is why only the Chebychev characteristic is introduced. A detailed discussion of the binomial characteristic can be found in [38]. The larger bandwidth of the Chebychev transformer is accompanied by the drawback of ripples in the pass band, see figure 2.7b. These ripples and the bandwidth cannot be adjusted separately from each other, but smaller ripples inevitably lead to a smaller bandwidth. The filter design is discussed extensively in [38] and [44], therefore only the important points are introduced here with an example of a two-stage Chebychev matching transformer.

After definition of the maximum allowable reflection in the passband $|\underline{S}_{11,\mathrm{max}}|$, an auxiliary value k can be calculated [44] by

$$k = \sqrt{\frac{|\underline{S}_{11,\mathrm{max}}|^2}{1 - |\underline{S}_{11,\mathrm{max}}^2|}}. \tag{2.46}$$

With the relation

$$\cot^4 \theta_z = \frac{4k^2 Z_L Z_0}{(Z_L - Z_0)^2}, \tag{2.47}$$

Table 2.3: Impedances of different sections in a Chebychev transformer along with the respective fractional bandwidths.

Stages	Z_1 (Ω)	Z_2 (Ω)	Z_3 (Ω)	Z_4 (Ω)	$\frac{\Delta v}{v_0}$ (%)
2	3.65	10.97	-	-	44
3	2.88	6.32	13.9	-	77
4	2.57	4.45	8.98	15.59	101

the characteristic impedances of the particular stages can be calculated by

$$Z_1^2 = Z_0^2 \sqrt{\frac{(Z_L - Z_0)^2}{4Z_0^2 \tan\theta_z} + \frac{Z_L}{Z_0} + \frac{(Z_L - Z_0)Z_0}{2\tan^2\theta_z}}, \text{ and} \tag{2.48}$$

$$Z_2 = \frac{Z_L}{Z_1} Z_0. \tag{2.49}$$

The characteristic impedance values for the particular transformer stages can also be obtained from tables (also given in [38]). The fractional bandwidth of a Chebychev transformer is then given by [38]

$$\frac{\Delta f}{f} = 2 - \frac{4\theta_m}{\pi}, \text{with} \tag{2.50}$$

$$\sec(\theta_m) = \cosh\left(\frac{1}{N}\cosh^{-1}\left(\frac{1}{|\underline{S}_{11,\max}|}\left|\frac{Z_L - Z_0}{Z_L + Z_0}\right|\right)\right). \tag{2.51}$$

In figure 2.7b, the $|\underline{S}_{11}|$ characteristics of Chebychev transformers are depicted for an impedance transformation from $Z_L = 20\,\Omega$ to $Z_g = 2\,\Omega$. This was calculated for a 2-stage, 3-stage and 4-stage transformer, assuming a maximum acceptable reflection $|\underline{S}_{11,\max}| = 0.05 = -26$ dB. Clearly visible are the ripples in the passband. Furthermore, already for $n = 2$, the fractional bandwidth is significantly increased as compared to the single stage transformer. The parameters of the separate sections are given in table 2.3 along with the respective fractional bandwidths.

Stubs

So far, the loads to the transmission lines were assumed arbitrary, yet not particularly discussing the extreme cases of real load impedances $\underline{Z}_L = 0$ or $\underline{Z}_L \to \infty$. This is schematically depicted in figure 2.8a. These cases are of particular interest especially for the tuning of complex loads.

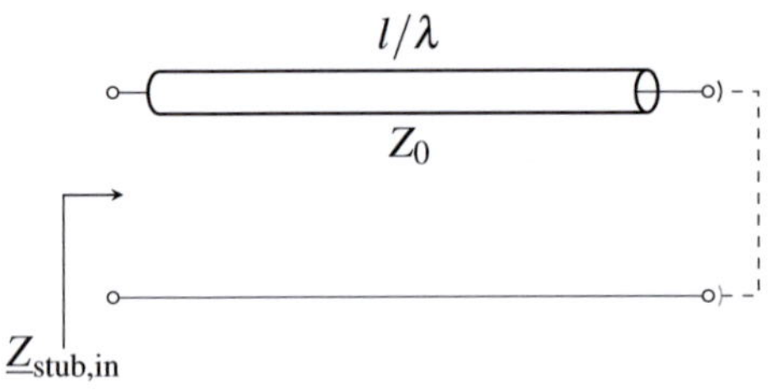

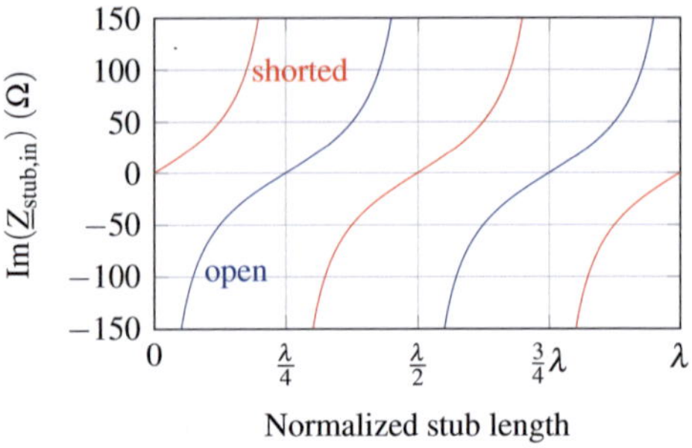

Figure 2.8: Equivalent circuit and input impedance of open ended stubs. The equivalent circuit of the stub connected to the termination is depicted in (a). The termination can be either an open end or a short circuit. This is depicted by the dashed line. The resulting input impedance of this stub transformation is depicted in (b), for the two cases of an open-ended stub and a shorted stub.

In these cases, equation (2.42) yields

$$\underline{Z}_{\text{in}} = \text{j}Z_0 \tan(\beta\ell), \quad \text{for } \underline{Z}_{\text{L}} = 0 \tag{2.52}$$

$$\underline{Z}_{\text{in}} = -\text{j}Z_0 \cot(\beta\ell), \quad \text{for } \underline{Z}_{\text{L}} \to \infty. \tag{2.53}$$

Obviously, the resulting input impedance is purely imaginary for any length ℓ and can take any value between $-\text{j}\infty$ and $+\text{j}\infty$. Figure 2.8 depicts the reactance of $\underline{Z}_{\text{in}}$ for transmission line lengths from 0 to $\ell = \lambda$. For $Z_{\text{L}} \to \infty$, starting from zero length, the resulting input impedance shows a capacitive behavior up to a length of $\lambda/4$. A shorted transmission line exhibits inductive behavior if it is shorter than $\lambda/4$.

This behavior enables the utilization of open-ended or short circuited transmission lines as tuning elements for complex loads. They can be connected in parallel or in series to a complex load, e.g. compensating the load susceptance, and are then called stubs.

Especially in the microstrip technology, a special case of the open-ended stub is the radial stub. The top view of such a structure is depicted schematically in figure 2.9a. The ground layer is hereby depicted in dark gray, whereas the microstrip structure and the radial stub are colored in lighter gray. The microstrip line coming from the left end of the image is in this case terminated by an arc shaped structure, the radial stub. The characteristic quantities of a radial stub are its radii $R_{\text{stub,o}}$ and $R_{\text{stub,i}}$ and the opening angle α_{Stub} as indicated in figure 2.9a. A detailed derivation of the properties of radial stubs can be found in [45–47]. From this, it can be seen that the bandwidth of the stub is essentially determined by α_{Stub} and the frequency influence is given by the stub radii, similar to the stub length of a straight stub. The radial stub can be connected to the transmission line in multiple ways.

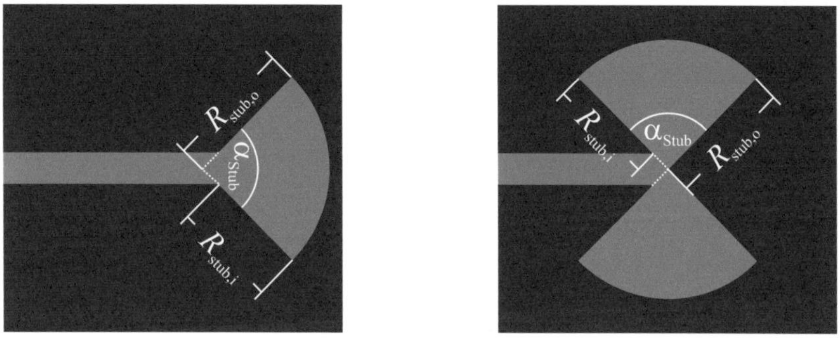

a) b)

Figure 2.9: Schematic top view on radial stubs in microstrip technology. The ground electrode is represented by the dark gray areas, whereas the microstrip line is depicted in light gray. Two flavors of radial stubs are depicted: the single radial stub (a) and the butterfly stub (b). Both stubs are connected to a microstrip transmission line pointing towards the left of each image.

As shown in figure 2.9a, in the propagation axis of the microstrip line, or perpendicular to the propagation direction. In the latter case, however, the opening angle is limited to less than $180°$. Increasing the opening angle can be done by mirroring the structure at the center, as shown in figure 2.9b. The opening angle is in this case doubled and due to its appearance, the structure is called butterfly stub.

The benefit of these radial stubs over the straight subs can be illustrated by the slope of the impedance as depicted in figure 2.10. The resistance of the radial stub exhibits a very low level across a large length range, also and especially at the point, where the reactance $\mathrm{Im}(\underline{Z}_{\mathrm{stub,in}})$ vanishes. For comparison purposes, the reactance of an open-ended straight stub as shown in 2.8 is depicted here as well. Apparently, the derivation of the reactance of the radial stub is much smaller, resulting in a larger bandwidth. Due to their low impedance, radial stubs are frequently used as wideband short-circuits for e.g. the termination of a microstrip line, without the need of a physical connection enabling e.g. the DC biasing of active devices. This is especially beneficial in the microstrip technology, since no interconnects through the dielectric are necessary. The schematics shown in figure 2.9 depict the case of a microstrip transmission line, terminated by a radial stub.

If the radial stub is designed properly, this entire structure can be used as a shorted stub and its impedance thus corresponds to the red lines in figure 2.8.

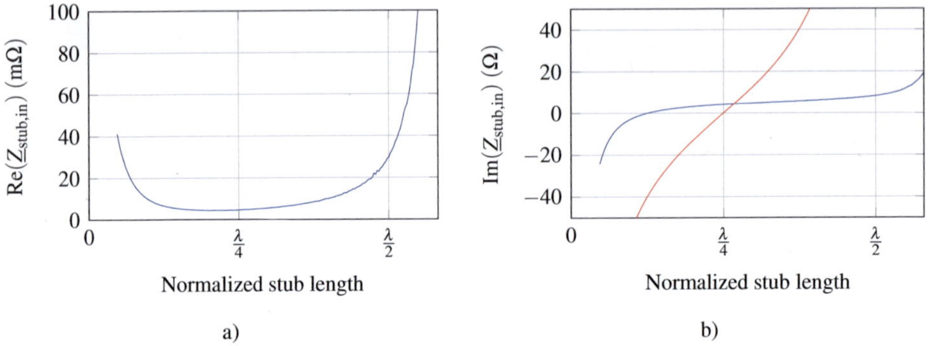

Figure 2.10: Comparison of the impedances of radial with straight stubs. (a) The real part of the input impedance of a radial stub, which exhibits a low resistance across a wide length range. The comparison of the resulting imaginary parts of the input impedance between a straight stub (red line) and a radial stub (blue line) is depicted in (b). Note the much shallower slope of the radial stub.

Conclusion

In this section, the concept of impedance matching by transmission lines is presented. If the load impedance is not matched to conjugate complex impedance of the generator, an impedance transformation is required, in order to achieve the maximum power transfer. Since the realization of discrete elements for impedance transformation is impractical in the sub-mm wavelength range, the impedance transformation is typically realized by transmission lines. The drawback of this approach usually is the limited bandwidth of the circuit. The most important transformation types are introduced for particular purposes. For the transformation of a real load, the $\lambda/4$ transformer can be used as single stage or multi-stage transformer. Tuning complex impedances can be achieved by the usage of either open-ended or short-circuited stubs. These components exhibit a solely imaginary impedance and can thus be applied to a complex load in order to achieve a real impedance.

2.3.4 Antennas

In order to efficiently couple a signal from free-space to a receiver, the front end of such a system requires the implementation of an antenna. In general, an antenna of a receiver is a component which transforms an electromagnetic wave propagating in free-space to a transmission line- or waveguide-bound wave. This includes the adjustment of the type of wave as well as an impedance transformation. Figure 2.11a depicts a simple schematic setup of a receiving antenna connected to a load through a transmission line. This feed line typically determines the physical setup of the antenna.

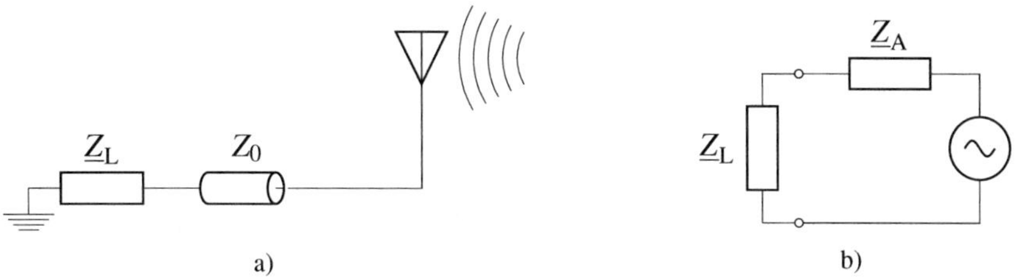

Figure 2.11: Equivalent circuit of an antenna. (a) An antenna connected to a load impedance through a transmission line. For modeling purposes, the antenna can therefore be regarded as a voltage source with a particular output impedance $\underline{Z}_A$ as depicted in (b).

Therefore, numerous kinds of antennas are available, designed to fit to the mechanical properties of their particular feed lines. In literature, there are various mixer systems that are based on waveguide coupled mixer systems with horn antennas [26, 48–50]. This work is however limited to the realization of planar structures, which is why the following section will only consider planar antennas, namely dipole and slot antennas. Since antennas are not subject of intense investigation and optimization, only the major antenna properties are shown in this section. Detailed analysis and derivations of the particular properties can be found in [51–56].

Efficient power transmission from free-space radiation to a load resistance requires the antenna to match the impedance of a plane wave in free-space

$$Z_0 = 120\pi\,\Omega \approx 377\,\Omega \tag{2.54}$$

to the characteristic impedance of the feed line. The antenna appears to the transmission line circuit as a component with a given impedance, the radiation impedance Z_A. A receiving antenna can therefore be modeled by the equivalent circuit given in figure 2.11b. At this point, it shall be noted, that due to the reciprocity theorem [38], an antenna has the same properties regardless of its operation mode. This implies that a given antenna exhibits the same properties for a receiving operation as well as for sending operation.

In order to characterize an antenna regarding its spatial dependence of the radiation power density P', the gain

$$G = \frac{P'_{\mathrm{max}}}{P'_{\mathrm{i}}} \tag{2.55}$$

is introduced, which correlates the power density in the antenna main lobe with the power density of an isotropic emitter. The radiation power density from this fictive construct in a distance r is given by

$$P'_{\mathrm{i}} = \frac{P_{\mathrm{S}}}{4\pi r^2}, \tag{2.56}$$

when the antenna is driven with the source power P_{S}. Due to the reciprocity, the gain applies to both, the sending and the receiving antenna. In case of a receiving antenna, this implies that the available power at the antenna connectors is altered by a factor of G compared to the isotropic emitter.

In the frame of this work, dipole antennas as well as slot antennas are used. These antennas can be realized in planar technology on a dielectric substrate. This dielectric introduces a spatial non-uniformity of the permittivity surrounding the antenna, which focuses the radiated power of the antenna into the substrate. In finite substrates, this leads to unwanted substrate modes [55]. This effect can on the other hand be exploited if the substrate is attached onto a lens that has a similar permittivity. This construct acts as a semi-infinite thick substrate with the power being radiated primarily into the dielectric, thus increasing the gain [55]. This setup comprising the dielectric substrate glued onto a dielectric lens is intended for the devices in this work.

For the design of an antenna, the geometrical as well as the electrical parameters need to be determined. In case of a $\lambda/2$ dipole antenna, the effective length needs to fit half a wavelength. The effective length also known as the electrical length is larger than the geometrical or physical length. A dipole antenna with a physical length of $\lambda/2$ exhibits a complex impedance and therefore, its length needs to be contracted with a factor of 0.95.

The second type of antennas that is used is the slot antenna, which is the dual type to the dipole antenna. Therefore, the Babinet's principle applies, which can be used to calculate the slot antenna radiation impedance Z_{S} by

$$Z_{\mathrm{S}} Z_{\mathrm{A}} = \frac{Z_0^2}{4}. \tag{2.57}$$

Due to their physical setup, it is convenient to use the slot antenna in coplanar waveguides and the dipole antenna in a microstrip transmission line environment, since the physical setup meets the layer setup of the transmission line technology. However, the antenna technology is not limited to the transmission line technology but can be interchanged.

Conclusion

This section gives a short introduction to antennas, which are essential for the radiation coupling from free space to a transmission line. The types of antenna used in this work are introduced, which are the dipole and the slot antenna. These antennas are dual antennas, which is why their characteristics are connected by the Babinet's principle. In order to achieve a maximized power transmission from an antenna to a load, it needs to be ensured, that the characteristic impedances are matched. This needs to be considered during antenna design, but can also be achieved by a subsequent impedance transformation. The antenna can in this context be regarded as the signal generator and in the frame of this work, the mixer acts as a load. Therefore, the transmission line network needs to account for the impedance matching of these two devices. The noise temperature of the mixer is of course dependent on this periphery, which is why a maximized power transmission from the antenna to the mixer has to be achieved. This does not only include the impedance matching in the RF circuit, but also the losses on the transmission lines. Therefore, superconducting transmission lines can be advantageous, which is why the next chapter is dedicated to the derivation of topic of superconductivity. Furthermore, the active devices required for the operation of an integrated receiver will be presented.

3 The superconducting integrated receiver

The first superconducting integrated receiver has been demonstrated by Koshelets et al. [57] and has been continuously developed ever since. It has already been employed in missions, such as in TELIS [58]. The relevant fundamentals of superconductivity and the Josephson effect are introduced in this chapter. Along with the phenomenologic and symptomatic behavior, the relevant theories are introduced, including the characteristic parameters. The superconductor - insulator - superconductor trilayer is subsequently introduced with its unique properties. Since the mixer junctions based on these SIS trilayers exhibit the same geometrical setup as Josephson junctions, the Josephson effect is introduced along with the equivalent circuit, which is the basis for the representation of these junctions in the high-frequency simulations. The behavior of a SIS junction with applied microwave radiation will be shown along with the derivation of the characteristic properties, defining the high frequency behavior. The Josephson effect and its characteristics is also of utmost importance for the development of the local oscillator, which is realized by a flux-flow oscillator and will be presented in the very end of the chapter.

3.1 Fundamentals of superconductivity

Ever since the discovery of superconductivity by Heike Kammerlingh Onnes in 1911 this topic has been subject to intense research. The superconducting transition occurs when the thermal energy of the electrons in a material undercut a certain threshold, which is defining the material dependent critical temperature T_c of a superconductor. At this temperature, the electrical DC-resistivity of the material sharply drops to an immeasurably small value. This behavior is schematically depicted in figure 3.1a by the blue line. The red line represents the $R(T)$ dependence of a normal metal which does not exhibit the superconducting transition. Figure 3.1b shows the measurements of the specific resistances ρ of different materials in a temperature range from room temperature to 4.2 K, in this case aluminum, niobium and niobium nitride.

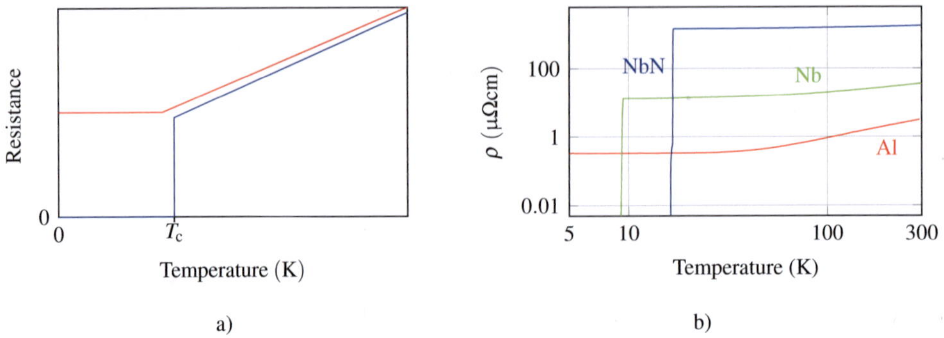

Figure 3.1: (a) The schematic resistance dependence of a metallic superconductor (blue line) and an example of a normal conducting material (red line). (b) Measured values of the specific resistance (ρ) dependence on the temperature for aluminum (red line), niobium and niobium nitride. Aluminum is not superconducting at temperatures above 1 K.

It is apparent, that aluminum does not undergo the superconducting transition yet, whereas the niobium exhibits a critical temperature at 9.2 K and the niobium nitride has an even higher transition temperature of 17 K. For temperatures $< T_c$, the superconducting state can only be observed, if the current density j_{SC} and the applied magnetic field H are below a threshold value which is called their critical values:

$$j_{SC} < j_{c,SC} \tag{3.1}$$

$$H < H_c. \tag{3.2}$$

In the superconducting state, the superconductor almost behaves as a perfect electrical conductor, with a slight but important difference: an applied magnetic field will be expelled from the material except for a thin surface layer, independent on the chronological sequence of cooling and application of the magnetic field. This makes the superconductor different from a perfect electrical conductor and the discovery of this Meissner- Ochsenfeld effect is the evidence for the superconducting state to be a separate thermodynamical state.

The first theory about the electro-dynamic behavior of superconductors was established by the London brothers in 1935. They proposed what came to be known as the London equations [59]

$$\vec{E} = \mu_0 \lambda_L^2 \frac{\partial}{\partial t} \vec{j}_{SC} \tag{3.3}$$

$$-\vec{B} = \mu_0 \lambda_L^2 \vec{\nabla} \times \vec{j}_{SC}, \tag{3.4}$$

connecting the electric and magnetic field strengths $\vec{E}$, $\vec{B}$ with the supercurrent density. The proportionality factor

$$\lambda_L = \sqrt{\frac{m_s}{\mu_0 q^2 n_s}} \tag{3.5}$$

is the London penetration depth with the vacuum permeability μ_0 and describes the length scale on which an applied magnetic field decays exponentially in the surface layer of the superconductor.

3.1.1 BCS Theory

It was not until 1957 that John Bardeen, Leon Neil Cooper and Robert Schrieffer introduced a microscopic theory of the superconducting state [60], the BCS theory. In this theory, Cooper pairs were proposed as the charge carriers in superconductors. This is based on the theory of an attractive force between two electrons in spite of their equal charge [61]. By interacting with the crystal lattice, two electrons with opposite spin and momentum can relax to a lower energy state, thus creating an energy gap $\Delta(T)$ in the superconductor around the Fermi level by forming a Cooper pair. In this bound state, the Cooper pairs can move along the crystal structure without scattering, resulting in the loss less current transport. The Cooper pairs can only be split into single electrons (called quasiparticles) again, if an energy larger than $2\Delta(T)$ is applied. As opposed to single electrons, the Cooper pairs have an integer spin and are thus not subject to the Pauli-principle disallowing the occupation of the same state of fermions. This boson-like behavior of Cooper pairs allows them to occupy the identical ground state and therefore, all Cooper pairs in a superconductor can be be described by a single, coherent wave function

$$\Psi(\vec{z},t) = \sqrt{n_S} \cdot e^{i\varphi_{SC}(\vec{z},t)} = |\Psi(\vec{z},t)|e^{i\varphi_{SC}(\vec{z},t)} \tag{3.6}$$

with the Cooper pair density n_S and the phase φ_{SC}.

The spatial extension of a Cooper pair is given by [62]

$$\xi_0 = \frac{\hbar v_F}{\pi \Delta(0)} \tag{3.7}$$

which is the material dependent BCS coherence length. $\hbar$ is the reduced Planck constant and v_F is the Fermi velocity.

The energy gap resulting from the condensation of electrons to Cooper pairs is quantified in this theory as

$$\Delta(0) = S k_B T_c \tag{3.8}$$

with the Boltzmann constant k_B and a parameter S, which was determined to $S = 1.764$ for weakly coupled superconductors. Superconductors in the strong coupling limit, such as niobium nitride, can exhibit values as high as 2.2 [63]. The energy gap is temperature dependent starting with a value of zero at $T = T_c$ and increasing towards lower temperatures according to

$$\Delta(T) = \Delta(0)\sqrt{\cos\left(\frac{\pi}{2}\left(\frac{T}{T_c}\right)^2\right)}. \tag{3.9}$$

For pure superconductors, the London penetration depth is smaller than the coherence length, which results in the Ginsburg-Landau parameter

$$\kappa = \frac{\lambda_{GL}}{\xi} \tag{3.10}$$

to be smaller than unity which was common at the time when this theory was published. However, it was published by Abrikosov [64] that superconducting materials with $\kappa > 1/\sqrt{2}$ exhibited a different behavior related to the breakdown of superconductivity induced by the magnetic field. These superconductors are called superconductors of the second kind. Similar to superconductors of the first kind, these materials are in the so-called Meissner state for $0 < H < H_{c1}$. Upon reaching the first critical magnetic field H_{c1}, the superconductivity does not break down abruptly, but decreases until it vanishes completely at a second critical field H_{c2}. For $H_{c1} < H < H_{,2}$ the superconductor is in the Shubnikov phase and can be penetrated by magnetic vortices which are called Abrikosov vortices. Each vortex carries a magnetic flux given by the magnetic flux quantum

$$\Phi_0 = \frac{h}{2e} = 2.0678 \cdot 10^{-15}\,\mathrm{Wb} \tag{3.11}$$

and exhibits a normal conducting core surrounded by screening currents. The second critical magnetic field can be derived with $B = \mu_0 \cdot H$ by

$$B_{c,2} = \frac{\Phi_0}{2\pi\xi_{GL}^2} \tag{3.12}$$

where ξ_{GL} denotes the Ginsburg-Landau coherence length of the superconducting wave function of a superconductor. This quantity describes the length scale on which the superconducting wave function can vary [62]. The coherence length of niobium $\xi_{GL,Nb} \approx 15$ nm [65, 66] is approximately four times larger than in niobium nitride $\xi_{GL,NbN} \approx 4$ nm [67].

The theories described above introduce the relevant parameters that are necessary in order to describe the superconducting devices that will be developed in this work.

3.1.2 Mattis-Bardeen theory

For DC currents, the superconductors are considered lossless. However, if RF signals are transferred, losses are present in superconductors. This section is dedicated to the derivation of the superconducting losses at RF frequencies, which is then modeled using the surface impedance concept.

This concept exploits the premise that for good conductors, an incident power is mostly reflected but the small amount of absorbed power is dissipated within a thin surface layer of the conductor [38]. The thickness of this layer in a normal metal is given by the skin depth

$$\delta_\mathrm{s} = \sqrt{\frac{2}{\omega\mu\sigma}}, \tag{3.13}$$

where $\sigma = 1/\rho$ represents the conductivity of the material and μ is the magnetic permeability. The surface impedance of a normal metal is given by [38]

$$Z_\mathrm{S,N} = \sqrt{\frac{\omega\mu}{2\sigma}} = \frac{1}{\sigma\delta}. \tag{3.14}$$

As opposed to DC operation, superconductors are not lossless at RF frequencies. In order to quantify these losses, a two-liquid model [62, 68] was proposed, establishing the idea that the current is carried by the coexisting quasiparticles as well as the Cooper pairs, leading to a complex conductivity.

However, this model neither includes effects close to the energy gap of the superconductor, nor the pair breaking by microwave radiation with frequencies above the gap frequency. Therefore, the more complex Mattis-Bardeen theory [69] is used which includes a complex conductivity

$$\sigma = \sigma_1 - j\sigma_2. \tag{3.15}$$

In superconductors, the imaginary part of the impedance is much larger than in normal metals, which is the reason for the calculation of the surface impedance in the anomalous limit.

The values of σ_1 and σ_2 can be determined numerically with

$$\begin{aligned}\frac{\sigma_1}{\sigma_n} = \frac{2}{\hbar\omega}\int_{\Delta}^{\infty}\frac{(f(E)-f(E+\hbar\omega))\left(E^2+\Delta^2+\hbar\omega E\right)}{\sqrt{E^2-\Delta^2}\sqrt{(E+\hbar\omega)^2-\Delta^2}}\mathrm{d}E+ \\ \frac{1}{\hbar\omega}\int_{\Delta-\hbar\omega}^{\Delta}\frac{(1-2f(E+\hbar\omega))(E^2+\Delta^2+\hbar\omega E}{\sqrt{E^2-\Delta^2}\sqrt{(E+\hbar\omega)^2-\Delta^2}}\mathrm{d}E\end{aligned} \quad (3.16)$$

$$\frac{\sigma_2}{\sigma_n} = \frac{1}{\hbar\omega}\int_{\max(\Delta-\hbar\omega,-\Delta)}^{\Delta}\frac{(1-2f(E+\hbar\omega))(E^2+\Delta^2+\hbar\omega E)}{\sqrt{E^2-\Delta^2}\sqrt{(E+\hbar\omega)^2-\Delta^2}\mathrm{d}E} \quad (3.17)$$

and the Fermi-Dirac distribution

$$f(E) = \frac{1}{\mathrm{e}^{\frac{E}{k_B T}}+1}. \quad (3.18)$$

σ_n represents the normal state conductivity of the material and can be calculated using

$$\sigma_n = \frac{\hbar}{\pi\Delta(0)\lambda_L^2(0)}. \quad (3.19)$$

The required quantities can be derived from temperature dependent resistance measurement around T_c. Δ can be obtained by measuring T_c and using (3.8) and (3.9). σ_n can be calculated by either (3.19) or simply by measuring $\rho_{T>T_c} = 1/\sigma_n$.

The surface resistance can then be determined with [44]

$$\begin{aligned}\underline{Z}_S = R_S + \mathrm{j}X_S &= \sqrt{\frac{\mathrm{j}\omega\mu_0}{\sigma}} + \coth\left(d\sqrt{\mathrm{j}\omega\mu_0\sigma}\right) \\ &= \sqrt{\frac{\mathrm{j}\omega\mu_0}{\sigma_1-\mathrm{j}\sigma_2}} + \coth\left(\frac{d}{\lambda_L}\sqrt{1+\mathrm{j}\frac{\sigma_1}{\sigma_2}}\right).\end{aligned} \quad (3.20)$$

Comparing the surface impedances of normal conductors to superconductors leads to the observation that superconductors exhibit a surface resistance for $T < T_c$ that is orders of magnitudes lower than in normal conductors, predestining the superconductors for the realization of low loss transmission lines.

Figure 3.2a shows the calculated surface resistance versus the frequency for different materials. The calculations are done assuming an operating temperature of 4.2 K and are based on temperature dependent resistance measurements of Nb, NbN and Al, which enabled the determination of T_c and σ_n as given in table 3.1. Since aluminum is not superconducting at that temperature, the formulas for a normal metal apply.

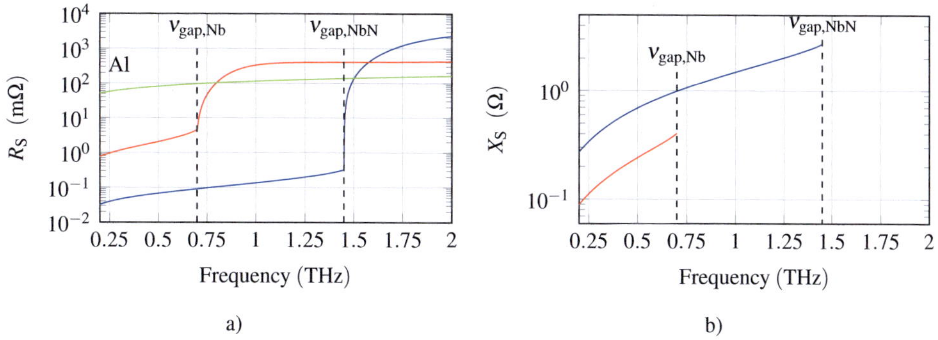

Figure 3.2: Surface impedances of niobium and niobium nitride, calculated using the Mattis-Bardeen theory. The real part of the impedance is depicted in (a), whereas the imaginary part is presented in (b). Clearly visible is the strong increase in resistance at the respective gap frequencies. The green curve represents the calculated surface resistance of aluminum. All calculations were performed for an operation temperature of 4.2 K.

Up to the so called gap frequency ν_{gap}, the surface resistances of Nb and NbN stay approximately two orders of magnitude lower than for aluminum. At the gap frequency the energy of photons equal the energy gap of the superconductor

$$h\nu_{gap} = 2\Delta(T), \tag{3.21}$$

which is why frequencies above the gap frequency can not be carried by Cooper pairs. Therefore, the losses increase dramatically above the gap frequency. This is especially important for the material choice for the realization of transmission lines, since the resistance is included in the transmission line model as R', see equation (2.13). For signal transmission at frequencies lower than the gap frequency, superconductors are preferable over normal metals. However, above the gap frequency, the normal metals may be preferable. In order to increase the low loss transmission to higher frequencies, the energy gap of the superconductors needs to be increased.

The imaginary part X_S of the surface impedance of a superconductor is depicted in figure 3.2b. It is attributed to the inertia of the Cooper pairs, which are traveling without scattering through the crystal. This is modeled as kinetic inductance L_{kin} [70] and adds to the geometric inductance of a transmission line (see figure 2.2) to $L' = L'_{geom} + L'_{kin}$. Obviously, the imaginary part outnumbers the real part of the impedance by orders of magnitudes.

Table 3.1: Measured parameters of different films for the determinations of the surface resistance.

Material	T_c (K)	ρ (µΩ cm)
Al	-	0.3
Nb	9.2	2.34
NbN	17	40

Conclusion

This section introduces the most important characteristics of superconductors. The fundamental theories are introduced giving an insight in the physical principles involved in superconducting materials. This includes the BCS theory which gives the theoretical explanation for the formation of an energy gap in a superconductor. For energies below this gap energy, the superconductor allows a lossless transportation of DC currents. However, RF currents are subject to losses in superconductors. The level of losses that has to be expected can be calculated using the Mattis-Bardeen theory. This theory is based on a complex conductivity in a superconductor which can be calculated by utilizing the surface impedance concept. The results from the calculation of the actual surface impedance values are complex values that account for the Ohmic losses in a superconductor and a certain imaginary part of the superconducting RF- impedance, which is given by the kinetic inductance of the superconducting material. Frequencies with energies below the superconducting gap can be carried by a superconductor with very small Ohmic losses. Therefore, the usage of superconductors as transmission lines guarantees a very high efficiency transmission line network, provided that the signal frequency is smaller than the gap frequency of the superconductor.

3.2 The tunnel junction

The unique properties of superconductors also enabled the path to active devices with unique properties. In 1962, Brian David Josephson postulated the tunneling of Cooper pairs between two superconductors through a thin, non-superconducting barrier without the dissipation of energy [71]. The experimental proof of this theory was already published one year later by Anderson and Rowell [72] and motivated an entire field of research until today. This section will introduce the Josephson effect with its properties and derive the device characteristics that are necessary for the realization of an integrated receiver.

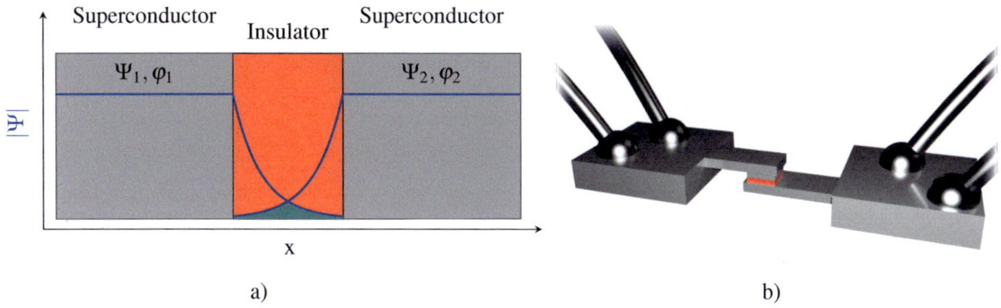

Figure 3.3: Schematic setup of a Josephson junction. (a) The interface of two superconductors separated by a thin insulating barrier. In the insulator, the wavefunctions Ψ_i overlap in spite of the exponential decay as indicated by the green area. (b) A more realistic setup of a Josephson junction yet not to scale. The superconductors are separated by the barrier, which is realized as layer on top of the ground electrode.

3.2.1 Josephson effect and short Josephson junctions

Josephson postulated the tunneling of Cooper pairs through a thin barrier due to the overlapping of the wave functions Ψ_1 and Ψ_2 of two superconductors. This barrier can be realized using normal metals (SNS) [73, 74], geometrically constricted superconductors [75, 76] or even insulators (SIS). In this work, only the insulating barriers will be considered. In an insulating barrier, the wave functions decay exponentially. Therefore, overlapping of the wave functions can only be achieved, if the barrier is much thinner than the coherence length of the superconductors. This is depicted schematically in figure 3.3a. Since the coherence lengths of the materials addressed here is only of the order of a few nm (see section 3.1.1), there is no technological process known that allows the lateral realization of the barrier (as suggested by figure 3.3a). Therefore, the barrier is typically realized as a layer in between two superconducting layers as shown in figure 3.3b. In this figure, the superconductor material is depicted in gray, the barrier is illustrated in red. Note that this figure is not to scale. The layer thickness can be controlled well enough in order to realize layer thicknesses of the order of only a few nm in a reproducible manner. The resulting Josephson junctions (JJs) can electrically be described by the first and second Josephson equation. These equations can be derived from the Schrödinger equation and can be found in detail in [77]. Assuming the absence of magnetic field, the Josephson equations are given by

$$I_\mathrm{S} = I_\mathrm{c} \cdot \sin(\varphi) \tag{3.22}$$

$$\frac{\partial \varphi}{\partial t} = \frac{2 \cdot \pi}{\Phi_0} \cdot V. \tag{3.23}$$

The first Josephson equation links the superconducting current I_S to the phase difference $\varphi = \varphi_2 - \varphi_1$ with a maximum current, which is the critical current I_c. If the restriction to zero magnetic field is not valid, the magnetic vector potential $\vec{A}$ has to be considered as well which leads to the gauge-invariant phase difference in a Josephson junction:

$$\varphi = \varphi_2 - \varphi_1 - \frac{2\pi}{\Phi_0} \int_{SC1}^{SC2} \vec{A} d\vec{s}. \tag{3.24}$$

The second Josephson equation relates the temporal phase differentiation to the voltage across a Josephson junction. The proportionality factor between the differentiated phase and the voltage is defined by physical constants exclusively. Assuming a constant voltage, the integration of (3.23) and subsequent substitution into (3.22) leads to the relation

$$I_s = I_c \left(\frac{2\pi}{\Phi_0} V t \right). \tag{3.25}$$

Since the voltage was assumed constant, the current obviously oscillates for a voltage $V \neq 0$. The frequency of this oscillation is given by

$$\nu_J = \frac{1}{\Phi_0} V = 483 \, \frac{\text{GHz}}{\text{mV}} \cdot V \tag{3.26}$$

and is called the Josephson frequency. Applying a voltage of 1 mV to the junction results in the oscillation of the current with 483 GHz. The inverse setup can be used by irradiating the junction with an incident signal frequency of 483 GHz in order to obtain an accurate output voltage of 1 mV.

The frequency and the voltage being only related via physical constants, is the reason for the accurate voltage standardization using Josephson junctions [78].

In order to model the behavior of a Josephson junction, the resistively and capacitively shunted junction (RCSJ) model [79] can be used, where we assume the junction to be small, i.e. the phase cannot vary across the junction. In this case the total current through a junction (I_{total}) is subdivided into three different components, a superconducting component I_S, a resistive component I_R and a displacement component I_D. The resistor models the normal state resistance of the junction, whereas the capacitor represents the junction intrinsic capacitance that is due to the parallel plate capacitor like setup of the Josephson junction. Figure 3.4 de-

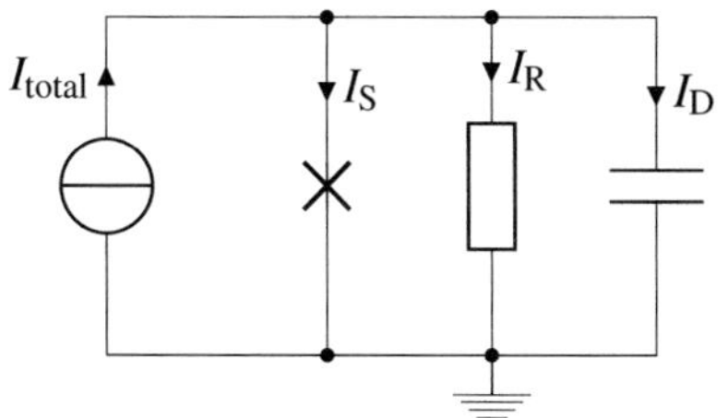

Figure 3.4: Equivalent circuit of a Josephson junction in the RCSJ model. The total current through the junction is subdivided into three different branches: the Josephson current I_S, the resistive current I_R and the current through the capacitor I_d.

picts the resulting equivalent circuit. The particular currents are given by the first Josephson equation and

$$I_R = \frac{V}{R_N} \tag{3.27}$$

$$I_D = C\frac{\partial V}{\partial t}. \tag{3.28}$$

Kirchhoffs law yields

$$I_{\text{total}} = I_S + I_R + I_D = I_c \sin(\varphi) + \frac{V}{R_N} + C\frac{\partial V}{\partial t}. \tag{3.29}$$

Applying the second Josephson equation results in a non-linear differential equation

$$I_{\text{total}} = I_c \sin(\varphi) + \frac{\Phi_0}{2\pi R}\frac{\partial \varphi}{\partial t} + C\frac{\Phi_0}{2\pi}\frac{\partial^2 \varphi}{\partial t^2}. \tag{3.30}$$

In order to illustrate the dynamics of the junction, a mechanical analog of a particle in sinusoidal potential is frequently used. Due to the appearance of this potential, the model is also called tilted washboard model. The differential equation, describing the dynamics of such a particle, is given by

$$0 = m\frac{\partial^2 x}{\partial t^2} + \nu\frac{\partial x}{\partial t} + \frac{\partial E_{\text{pot}}(x)}{\partial x}. \tag{3.31}$$

Comparing this equation with (3.30) leads to the analogies given in table 3.2. In this analogy, zero current corresponds to a horizontal sinusoidal potential and the particle is trapped in a potential well. It can only oscillate with the plasma frequency

$$\omega_{\text{plasma}} = \sqrt{\frac{2\pi I_c}{\Phi_0 C}} \tag{3.32}$$

Table 3.2: Analogies between the variables of a Josephson junction and a tilted washboard

Particle		Josephson junction	
location	x	φ	phase
speed	v	$\frac{\partial \varphi}{\partial t} = \frac{2\pi}{\Phi_0}V$	voltage
mass	m	$\frac{\Phi_0}{2\pi}C$	junction capacitance
friction	η	$\frac{\Phi_0}{2\pi}\frac{1}{R}$	conductance
driving force	$-\frac{\partial E(\varphi)}{\partial \varphi}$	I	current

around its ground state. This is illustrated in figure 3.5 by the red curve. Applying a current to the Josephson junction corresponds to a tilting of the potential. As long as the current does not reach the critical value, the particle will stay in the same well. However, if a current $\geq I_c$ is applied, the particle can travel down the potential. In this case, the mean velocity of the particle is not zero any more, corresponding to a voltage across the Josephson junction. If the current is reduced again, the mass of the particle along with the friction coefficient are determining the behavior of the motion of the particle. In the case of a large mass and a low friction, the current has to be reduced far below I_c in order to stop the unidirectional motion of the particle again and trap it in a potential well. The voltage and the current therefore exhibit a hysteresis. The large mass and the small friction in the analogy correspond to a large capacity and resistance respectively. The opposite case, when the particle has low mass and a high friction, the particle can get trapped easily when the current is reduced to I_c. In this case, the current-voltage characteristic (IVC) is non-hysteretic. In order to quantify the damping level of a junction, the Stewart-McCumber parameter

$$\beta_C = \frac{2\pi I_c R^2 C}{\Phi_0} \tag{3.33}$$

can be used. For $\beta_C \gg 1$, the junction exhibits a hysteresis, whereas $\beta_C \ll 1$ results in a non-hysteretic behavior which is desired in some applications such as SQUIDs or voltage standards.

In the case of SIS mixers, a highly hysteretic current-voltage characteristic as depicted in 3.6a is required. Josephson junctions with an insulating barrier intrinsically exhibit such a hysteretic IVC. This figure depicts an ideal current-voltage characteristic. By means of this example, the characteristic electrical parameters of a Josephson junction are introduced. Increasing the current starting from zero, the junction is in the zero voltage state which corre-

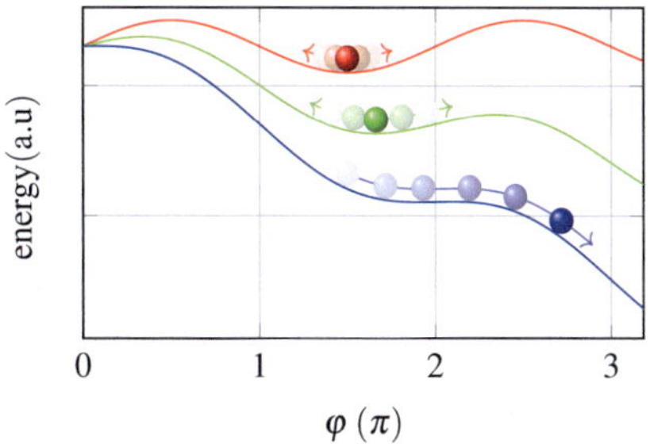

Figure 3.5: Tilted washboard potential model. The tilting sinusoidal potential is indicated by the three different curves. Exceeding a threshold value of the tilting angle results in a continuous movement of the particle in this potential as indicated by the blue curve.

sponds to the lossless Cooper pair tunneling through the barrier. The current can be increased up to a critical value I_c after which a voltage of

$$V_{gap} = \frac{2\Delta}{e} \tag{3.34}$$

appears. This voltage is thus called the gap voltage. The critical current I_c is defined by the critical current density j_c and the area A of the junction

$$I_c = j_c A. \tag{3.35}$$

The critical current density is anti-proportional to the barrier thickness which means that thinner barriers lead to larger critical current densities. In order to qualify the Cooper-pair tunneling properties of the barrier, the Ambegaokar-Baratoff parameter was introduced [80]

$$I_c R_N = \frac{\pi}{2e}\Delta(T)\tanh\left(\frac{\Delta(T)}{2k_B T}\right), \tag{3.36}$$

which assumes ideal tunneling of the Cooper pairs and is only dependent on the energy gap of the superconductors since $I_c \propto A$ and $R_N \propto \frac{1}{A}$. R_N hereby represents the normal state resistance of the junction. For $I > I_c$, the junction is in the voltage state, where the current transport is due to quasiparticle tunneling.

A further increase in the current at first does not further increase the voltage, but the junction stays at the gap voltage with a zero differential resistance due to the infinitely high density of states right at the edge of the gap which requires a perfect interface from the superconductors to the barrier [81]. After exceeding a threshold which is given by approximately V_{gap}/R_N, the junction is driven into its normal state, with the normal state resistance R_N. If the current is

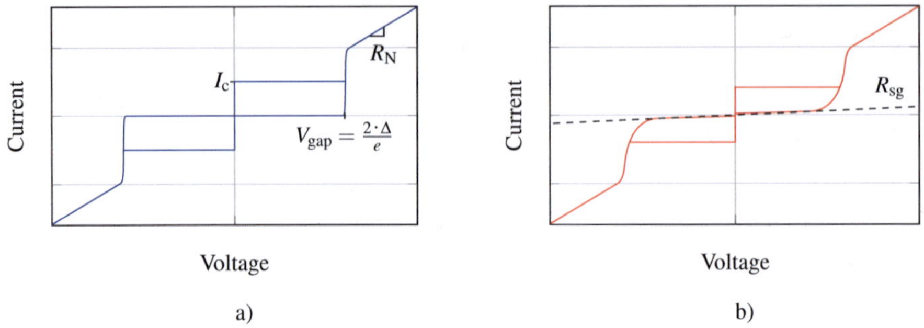

Figure 3.6: (a) The current-voltage characteristic of an ideal junction. The parameters I_c, V_{gap} and R_N are indicated in the graph. (b) A schematic illustration of a more realistic behavior in order to illustrate the difference to an ideal junction. The most prominent different characteristics are the decreased subgap resistance R_{sg} and the gap smear.

reduced, the voltage state is not reset at the critical current, but shows the expected hysteresis for SIS junctions. In an ideal junction, the barrier exhibits no leakage, and the current thus has to be set to almost zero in order to reset the junction to the zero voltage state. The small amount of current is due to a current given by thermally excited quasiparticles due to the finite operation temperature. However, at the typical operation temperature of 4.2 K, this can be neglected, especially since the influence of imperfections in the junction usually dominate the quality deterioration.

In order to quantify the quality of a real junction with respect to an ideal one, the quality parameters mentioned above can be used. However, in a real junction, there are some further features that have to be considered. Figure 3.6b schematically depicts a current-voltage characteristic of a real junction. This "non-ideal" Josephson junction exhibits a less sharp current rise at the gap voltage, which is often referred to as "gap smear", and is due to free states at the interface of the superconductors to the barrier [82]. Additionally, the retrapping current is not at the low value as depicted in 3.6a, but significantly increased. This increased current is mostly due to imperfections in the barrier resulting in leakage currents. Also present in many Josephson junctions (but not visible in figure 3.6b) is the presence of further sub-gap structures in the current-voltage characteristic [81]. These structures can be due to multiple Andreev reflections (MAR) [83] and increase the current in the sub-gap branch.
In order to quantify the sub-gap currents, the resistance R_{sg} can be utilized. This resistance is obtained at a voltage of $0.7 \cdot V_{gap}$. In order to obtain an area independent quantity, this sub-gap resistance is normalized to the normal state resistance R_N.

There are thus several parameters that are required to evaluate the quality of a Josephson junction. The gap voltage is given by the energy gaps of the applied superconductors. There-

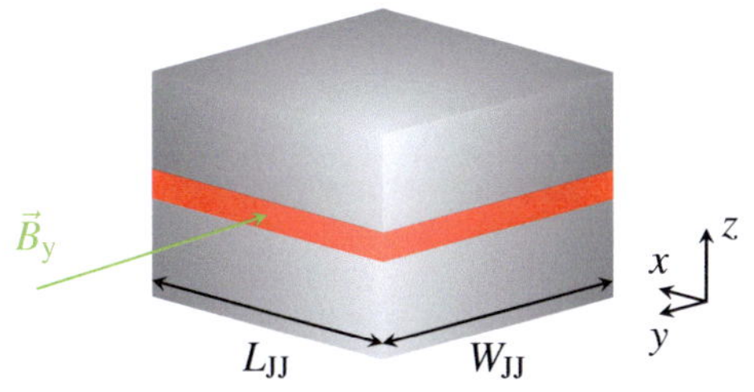

Figure 3.7: Schematic cross section of a Josephson junction applied with a magnetic field. The superconductors are depicted in gray, the barrier is colored in black. The magnetic field applied in plane with the barrier can penetrate the junction through the insulating barrier.

fore, the energy gap is directly related to Δ of the superconducting films that are used. In SIS-based mixers, the energy gap is the limiting factor concerning the operation frequency. Therefore, increasing the operation frequency of a SIS mixer requires the increase of the energy gap in the superconducting electrodes. Furthermore, imperfections in the barrier and its vicinity need to be avoided in order to obtain minimum leakage current and maximized non-linearity of the SIS junction.

So far, any magnetic field was considered absent. However, in case of a present magnetic field, the critical current of the junction is influenced according to (3.22) and (3.24). A magnetic field applied in plane with the barrier of the junction can penetrate not only the insulating barrier t_{ox} but also the superconducting electrodes down to a depth of λ_{L}. This is depicted schematically in figure 3.7. Therefore, the physical barrier thickness t_{ox} is different from the magnetic barrier thickness

$$t_{\text{ox,m}} = t_{\text{ox}} + \lambda_{\text{L},1} \coth\left(\frac{t_1}{\lambda_{\text{L},1}}\right) + \lambda_{\text{L},2} \coth\left(\frac{t_2}{\lambda_{\text{L},2}}\right) \tag{3.37}$$

where the thickness of the two electrodes is denoted by $t_{1,2}$. A junction which is aligned to a magnetic field perpendicular to its length L_{JJ} is thus penetrated by the magnetic flux

$$\Phi = B_{\text{y}} L_{\text{JJ}} t_{\text{ox,m}} \text{d}x, \tag{3.38}$$

as shown in figure 3.7.

According to equation 3.24, an applied magnetic field results in spatial variation of the phase difference. Integration of the magnetic field across the junction yields

$$\frac{2\pi\Phi}{\Phi_0} = \varphi(x + \text{d}x) - \varphi(x), \tag{3.39}$$

which can be expressed for a homogeneous magnetic field distribution as

$$\frac{\varphi(x+\mathrm{d}x)-\varphi(x)}{\mathrm{d}x} = \frac{\partial\varphi}{\partial x} = \frac{2\pi t_{\mathrm{ox,m}}}{\Phi_0} B_{\mathrm{y}}. \tag{3.40}$$

Substituting into equation 3.22, this results in a sinosuoidally oscillating current density within the junction

$$j(x,y) = j_{\mathrm{c}} \sin\left(\frac{2\pi}{\Phi_0} B_{\mathrm{y}} x + \varphi_0\right). \tag{3.41}$$

The oscillation of the current density in the junction is equivalent to a circulating current. It can be shown [62, 81], that the critical current of a short, rectangular junction which is perfectly aligned with the magnetic field, depends on the magnetic flux as

$$I_{\mathrm{c}}(\Phi) = I_{\mathrm{c}}(0) \left| \frac{\sin\left(\pi \frac{\Phi}{\Phi_0}\right)}{\pi \frac{\Phi}{\Phi_0}} \right|. \tag{3.42}$$

This function is the same as the intensity distribution of light diffracted by a single slit, and in this analogy, the ideal modulation of the critical current in a short junction by a magnetic field is called the Fraunhofer pattern. This result is only obtained, if the critical current density across the rectangular junction is constant, so the current distribution across the junction is homogeneous. In this case, the critical current distribution can be assumed as shown in figure 3.8a. This critical current modulation starts from $I_{\mathrm{c},0}$ for zero magnetic field and is decreased for increased $\left|\vec{B}\right|$ until it reaches zero. At this point, the flux in the junction equals $\Phi = \Phi_0$ and is called a Josephson vortex or a fluxon. For integer values of $n \cdot \Phi_0$, these distinct minima with a critical current of zero are present. The resulting Fraunhofer pattern is shown in figure 3.8b. The critical current density distribution can thus be evaluated using the modulation of the critical current due to a magnetic field.

Conclusion

In this section, the Josephson effect is introduced, with the particular properties that stem from its unique setup. The Josephson equations are presented which are the basis for the unique property of frequency to voltage conversion, with only fundamental physical constants as the proportionalities. The basic model of such a junction, the RCSJ-model and a mechanical analogue is introduced: the tilted washboard potential. Using this analogue, the dynamic behavior of a short Josephson junction can be explained. The equivalent circuit of the RCSJ-model is the basis for further modeling of a junction in a RF environment. The

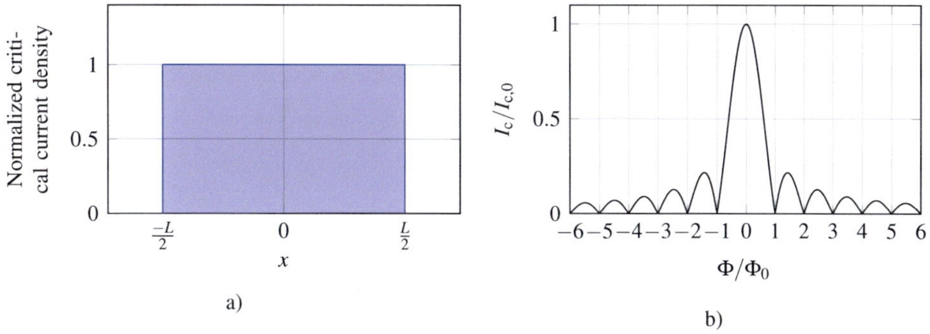

Figure 3.8: Modulation of the Josephson critical current of a junction with a magnetic field. These considerations are based on a perfectly square shaped junction with an assumed ideal critical current density distribution as depicted in (a) and a magnetic field in plane with the barrier. The applied magnetic field then modulates the Josephson critical current with a Fraunhofer pattern as depicted in (b).

quality parameters relevant for this work are introduced by the means of the current-voltage characteristic of an ideal junction as well as a non-ideal junction. It can be seen from these graphs, that the junction current-voltage characteristic is highly non-linear, which enables the usage as a mixing element.

3.2.2 SIS junction as mixer

As seen in section 2.2, a device for mixing needs to exhibit a non-linearity. In general, quite a few devices with non-linear transmission characteristics are known that can be used for mixing, such as Schottky diodes, hot electron bolometers (HEB) or SIS junctions. The Schottky diode however is outperformed by both the HEB as well as the SIS mixer [9]. The best performance of these three devices is achieved by the SIS mixer, closely followed by the HEB mixer. The HEB mixer, typically consisting of a thin niobium or niobium nitride film, exhibits a strong non-linearity in the current-voltage characteristic at the critical current of the thin film. It does not exhibit an intrinsic frequency limitation which makes it a common choice for frequencies that can not be covered by the SIS mixers. Furthermore, the HEB mixer exhibits a very low intrinsic capacitance, which facilitates the impedance matching even at high frequencies. The operation principle and properties of the SIS mixers, which are the devices with the best noise performance, will be discussed in this section.

Obviously, a Josephson junction exhibits strong non-linearities at $I = I_c$ and in the sub-gap branch where $I \approx 0$. Both of these non-linearities can be used as mixing [2, 16]. If the transition from the zero-voltage state to the voltage state is used for mixers, the device is referred to as Josephson-mixer.

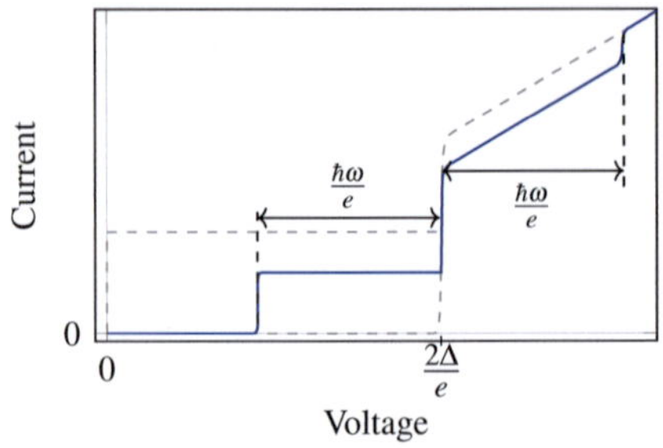

Figure 3.9: Current-voltage characteristic of a Josephson junction with incident RF radiation. The original characteristic is depicted by the dashed gray lines. The incident radiation with a frequency ν leads to photon assisted tunneling steps in the quasiparticle branch. The width of these steps is determined by the frequency of the incident radiation.

Although the transition at $I = I_c$ is extremely sharp, the performance of Josephson mixers is worse than of SIS-mixer, which uses the second non-linearity in the sub-gap branch. [2, 16]. Therefore, the Josephson mixers are disregarded in this work. In order to avoid additional noise by the Josephson effect [84], the critical current of the junction is usually suppressed using a magnetic field, leaving only the quasiparticle branch of the IV characteristic. All mixer devices referred to from now on, are SIS mixers.

In order to qualify a mixer for RF operation, the RF behavior of these devices needs to be considered. If a junction is provided with RF power, in the analogy of the RCSJ model, the incident RF current periodically tilts the washboard potential with the frequency of the incident signal. This periodic tilting can lead to resonance effects and result in a voltage that is proportional to the incident frequency. In the current-voltage characteristic, these resonances manifest themselves as deviations from the zero voltage state resulting in very sharp steps of constant voltage and are called Shapiro steps. The discrete voltages of these steps can be calculated directly from the Josephson equations and appear at the values

$$V_{\text{Shap}} = n\frac{\hbar\omega}{2e}. \tag{3.43}$$

The RF-power also manifests itself in the quasiparticle branch as photon-assisted tunneling steps (from now on called photon steps) as depicted in figure 3.9. The original IV-curve without incident RF-power is depicted as the gray dashed line, whereas the blue curve represents the behavior in the retrapping branch with incident RF-power. The width of these steps is given by

$$V_{\text{PAT}} = n\frac{\hbar\omega}{e}, \tag{3.44}$$

and is thus only dependent on the incident frequency. The integer n implies that these photon steps can occur at integer multiples of $\hbar\omega/e$. In figure 3.9 only the first photon step is depicted. Clearly visible is the deviation of the blue curve from the original one below as well as above the gap voltage. The reason for this can be illustrated by the energy diagram of the SIS junction as shown in figure 3.10. This diagram shows the occupied and the unoccupied states of the two superconducting electrodes separated by the barrier. In this barrier, there are no states available. The dashed horizontal lines depict the Fermi levels for each superconductor whereas the gray areas represent the occupied states. For a given bias voltage $0 < V_0 < 2\Delta/e$, an incident photon with sufficient energy $E = \hbar\omega$ enables a quasiparticle from the one electrode to overcome the energy difference to an unoccupied state in the opposite superconductor. This results in an enhancement of the current below the gap voltage, which can be measured in the current-voltage characteristic. A larger incident power hence leads to a higher current in the sub-gap branch. This has been formulated in [16, 85]

$$\Delta I_{\mathrm{qp}}(V_0, V_{\mathrm{LO}}) = \sum_{-\infty}^{+\infty} J_{\mathrm{n}}^2 \left(\frac{eV_{\mathrm{LO}}}{\hbar\omega}\right) I_0 \left(V_0 + \frac{n\hbar\omega}{e}\right), \tag{3.45}$$

where ΔI_{qp} represents the current rise on the nth photon step, V_{LO} is the amplitude of the incident local oscillator voltage and $I_{\mathrm{dc}}(V)$ is the current of the unpumped DC current-voltage characteristic with the bias voltage V_0. J_{n} is the Bessel function of the first kind of the order n. This can be approximated by [16]

$$\Delta I_{\mathrm{qp}} = \frac{1}{4} V_{\mathrm{LO}} \left(\frac{I_{\mathrm{dc}}(V_0 + \hbar\omega/e) - 2I_{\mathrm{dc}}(V_0) + I_{\mathrm{dc}}(V_0 - \hbar\omega/e)}{(\hbar\omega/e)^2}\right). \tag{3.46}$$

In order to quantify the pumping level of the mixer, a pumping factor α_{SIS} is introduced

$$\alpha_{\mathrm{SIS}} = \frac{eV_{\mathrm{LO}}}{\hbar\omega}. \tag{3.47}$$

With the formulas given in [16] the incident power to a SIS mixer can be calculated numerically. An approximation of the required power level for the ideal pumping is given by [84]

$$P_{\mathrm{LO}} = \frac{V_{\mathrm{LO}}^2}{2R_{\mathrm{N}}}. \tag{3.48}$$

From equations (3.47) and (3.48), it is obvious that the required pumping power increases with ω^2. Instead of just using a single SIS junction as the mixing element, also N junctions can be used. The required pumping power then increases by N^2 which includes an increase in saturation power by N^2 without deteriorating the mixer noise or gain [86]. It is shown that

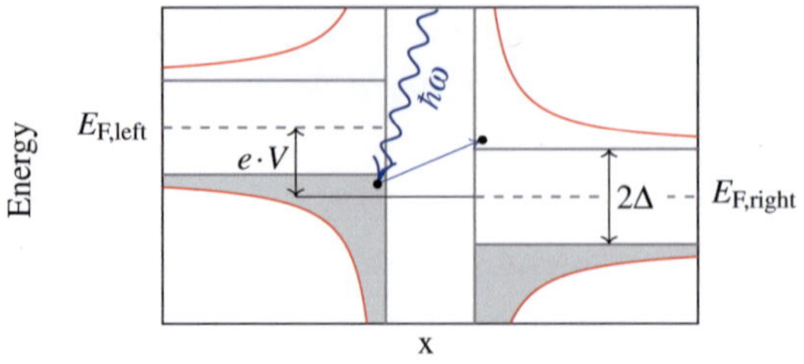

Figure 3.10: Energy diagram of a Josephson junction demonstrating photon assisted tunneling. The occupied states of each superconductor to the left and the right of the barrier are depicted in gray, below the energy gap, which surrounds the respective Fermi levels.

the photon steps can exhibit a negative differential resistance on the plateau. In this special case a SIS mixer can actually exhibit a conversion efficiency greater than unity which results in a conversion gain [16, 39, 41, 87].

Considering the width of the photon-assisted tunneling steps, the frequency limitation of the SIS mixer becomes obvious. The non-linearity will only be completely available to the RF-power if the energy of the incident radiation is lower than the energy gap 2Δ

$$E_{\mathrm{RF}} = \hbar\omega \leq 2\Delta = E_{\mathrm{gap}}. \tag{3.49}$$

A SIS mixer subjected to a frequency exceeding this limit, will exhibit a decreased conversion gain. However, it has been shown [88–91], that the SIS mixers still exhibit excellent performance until theoretically twice the energy gap. Therefore, the frequency limitation of a SIS mixer is given by

$$\nu_{\mathrm{mix,max}} = 4\Delta. \tag{3.50}$$

In case of niobium, which exhibits an energy gap of $2\Delta = 2.9$ meV, the SIS mixers can operate with acceptable noise levels up to $\nu = 1.4$ THz. In order to increase this limit, superconductors with higher energy gaps have to be used, such as niobium nitride.

The high frequency behavior can be modeled using the equivalent circuit of the RCSJ model. Since the critical current is suppressed, the Josephson component is removed from the circuit, leaving a resistor in parallel with a capacity as shown in figure 3.11a. The knowledge of the complex mixer impedance is essential in order to be able to match the RF circuit. Operating the mixer on the first photon step, the mixer conductance is described by Tuckers quantum theory as [16, 25]

$$G_{\mathrm{mix}} = \frac{e}{2\hbar\omega}\sum_{n} J_{\mathrm{n}}^{2}(\alpha_{\mathrm{SIS}})\left(I_{\mathrm{dc}}\left(V_0 + \frac{(n+1)\hbar\omega}{e}\right) - I_{\mathrm{dc}}\left(V_0 + \frac{(n-1)\hbar\omega}{e}\right)\right), \tag{3.51}$$

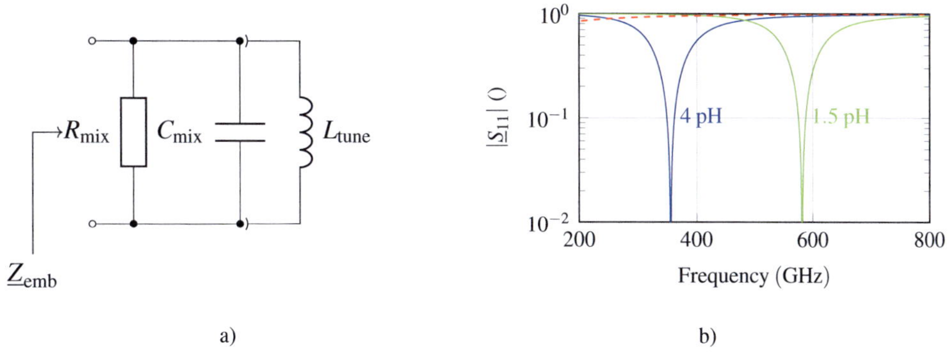

Figure 3.11: The equivalent circuit of a SIS mixer as derived from the RCSJ model, with suppressed Josephson effect. The remaining resistor and capacitor define the mixer impedance as depicted in (a) with a connected tuning inductor. (b) The calculation of $|\underline{S}_{11}|$ of the circuit in (a) with a capacity of $C_{\text{mix}} = 32$ fF. The uncompensated case is depicted by the dashed line. The addition of different inductors leads to the curves depicted in green and blue.

where J_n denotes the Bessel function of the first kind, V_0 is the bias voltage and I_{dc} is the current of the junction without incident radiation. It is shown by Tuckers quantum theory that SIS mixer exhibit a quantum RF susceptance [16, 25], which in the equivalent circuit is in parallel to the geometric capacitance $C_{\text{mix}} = C_{\text{geom}} + C_{\text{quantum}}$. However, this susceptance can be neglected if the junction is biased on the first photon step [1]. Therefore, the total susceptance is dominated by the geometrical capacitance. The input admittance of the junction is thus determined by the normal state resistance and the capacitance of the junction and equals

$$Y_{\text{in}} = G_{\text{mix}} + \mathrm{j}\omega C_{\text{mix}}. \tag{3.52}$$

In general, it has been shown [48, 92] that $R_{\text{mix}} = 1/G_{\text{mix}} = R_\text{N}$ is a good approximation for the mixer resistance. The mixer susceptance needs to be compensated for in order to achieve a good impedance matching which can be done by a variety of approaches some of which will be discussed in chapter 4. Here, the influence of tuning will be demonstrated with the simplest approach. In order to compensate for the junction capacitance, an inductive part has to be connected in parallel to the junction. This can be achieved by an ideal inductor. However, in order to allow the mixer to be dc biased, an additional capacitor would be required in series to the junction, which is very impractical at THz frequencies. Therefore, an approach using the impedance transformation of transmission lines is frequently used [92]. As derived in section 2.3.3, the input impedance of an open or short circuited transmission line exhibits a purely imaginary value. An open ended transmission line can exhibit a positive susceptance, if the length exceeds $l = \lambda/4$, whereas a shorted one acts inductively from $0 < l < \lambda/4$. Due to the shorter length, the bandwidth of a shorted transmission line is larger than an open circuited one. In order to provide an adequate RF-short, a radial stub can be used [92]. Figure 3.11a depicts the resulting circuit where the transmission line and the stub were substituted by an inductor in order to illustrate the operation principle being a parallel resonant circuit. The resulting embedding impedance is for an ideal circuit given by

$$\underline{Z}_{\mathrm{emb}} = \left(\frac{1}{R_{\mathrm{N}}} + \mathrm{j}\omega C_{\mathrm{mix}} + \frac{1}{\mathrm{j}\omega L_{\mathrm{tune}}}\right)^{-1}, \tag{3.53}$$

which exhibits a reactance of zero at the resonance frequency

$$\omega_{\mathrm{res}} = \frac{1}{\sqrt{L_{\mathrm{tune}} C_{\mathrm{mix}}}}. \tag{3.54}$$

The operation frequency can thus be selected by setting the length of the transmission line, which results in different values of the inductor. This is depicted in figure 3.11b for an uncompensated circuit and for compensated circuits with the values $L_{\mathrm{tune}} = 4\,\mathrm{pH}$ and 1.5 pH and $C_{\mathrm{mix}} = 32\,\mathrm{fF}$ resulting in resonance frequencies of $\nu_{\mathrm{res}} = \frac{\omega_{\mathrm{res}}}{2\pi} = 356\,\mathrm{GHz}$ and $\nu_{\mathrm{res}} = 581\,\mathrm{GHz}$ respectively. This small change in inductance of only 2.5 pH resulting in the large change in resonance frequency emphasizes the importance of an accurate tuning of the compensation. The uncompensated case is depicted by the red dashed line. Obviously, the reflection coefficient is close to unity over the entire frequency range, demonstrating the imperative need for this compensation.

The bandwidth of a resonant circuit can be calculated according to

$$BW = \frac{\Delta\nu}{\nu_{\mathrm{res}}} = \frac{1}{\omega_{\mathrm{res}} RC}. \tag{3.55}$$

Substituting the values for SIS mixers yields [93]

$$BW = \frac{1}{\omega_{res} R_{\mathrm{N}} C_{\mathrm{mix}}} = \frac{j_{\mathrm{c}}}{\omega_{res} I_{\mathrm{c}} R_{\mathrm{N}} C'_{\mathrm{tri}}}. \tag{3.56}$$

$C'_{\mathrm{tri}} = \frac{C_{\mathrm{mix}}}{A}$ hereby denotes the specific trilayer capacitance of the SIS trilayer used for fabrication. This value is material specific and can be found in literature for NbN/AlN/NbN trilayers [94] empirically determined to

$$C'_{\mathrm{tri}} = 20 j_{\mathrm{c}}^{0.6} \quad \text{for } j_{\mathrm{c}} < 15\,\mathrm{kA/cm^2} \tag{3.57}$$

$$C'_{\mathrm{tri}} = 71 j_{\mathrm{c}}^{0.16} \quad \text{for } j_{\mathrm{c}} > 15\,\mathrm{kA/cm^2}, \tag{3.58}$$

and for Nb/Al-AlO$_x$/Nb trilayers [95]

$$C'_{\mathrm{tri}} = 3.7 j_{\mathrm{c}} + 37, \tag{3.59}$$

where j_c is used in $\mathrm{kA/cm^2}$ and C'_{tri} in $\mathrm{fF/\mu m^2}$. An increase in j_c involves a decrease in the specific resistance of the trilayer $R'_{\mathrm{tri}} = R_N/A$. However, for a given increase in j_c, the decrease in R'_{tri} is larger than the increase in C'_{tri} [92, 95]. Assuming $C_{\mathrm{mix}} \propto A$ and $I_c \propto A$, equation 3.56 leads to the conclusion that an increased critical current density along with a decreased junction area results in a larger bandwidth [92, 96].

Conclusion

Due to the non-linearity of the current-voltage characteristic of a SIS junction, these devices are utilized as mixing elements. This is discussed in this section, with respect to both non-linearities: at I_c and in the sub-gap branch. Due to the increased noise levels, the junctions are typically not operated at the Josephson critical current non-linearity, but in the sub-gap branch. This section presented the influence of RF radiation incident to a Josephson junction and the effects on the current-voltage characteristic: the Shapiro steps and the photon-assisted tunneling steps. In order to realize a mixer from such a junction, it has to be embedded into a RF network. Therefore, the junction impedance has to be determined. The total susceptance is usually dominated by the geometrical capacity of the junction. The normal state resistance is typically a good approximation for the input resistance of a junction. For the realization of an integrated receiver, a local oscillator is required on chip. It is beneficial for the fabrication, if this local oscillator can be realized in the same technology as the mixer.

3.2.3 Long junctions and flux-flow oscillators

In the previous sections, it has been assumed, that the current distribution in the junction was spatially homogeneous. This assumption is valid for short junctions, which do not exceed a characteristic length scale

$$\lambda_J = \sqrt{\frac{\Phi_0}{2\pi\mu_0 t_{\mathrm{ox,m}} j_c}}. \tag{3.60}$$

λ_J is the so-called Josephson penetration depth. For these junctions, the phase is constant across the entire junction area in the absence of an external magnetic field. If the junction however is extended in one dimension, exceeding λ_J, the current distribution cannot be considered homogeneous any more, thus a phase change along the junction has to be taken into account. Therefore, a geometry as depicted in figure 3.12 is considered with a length L_{JJ} that

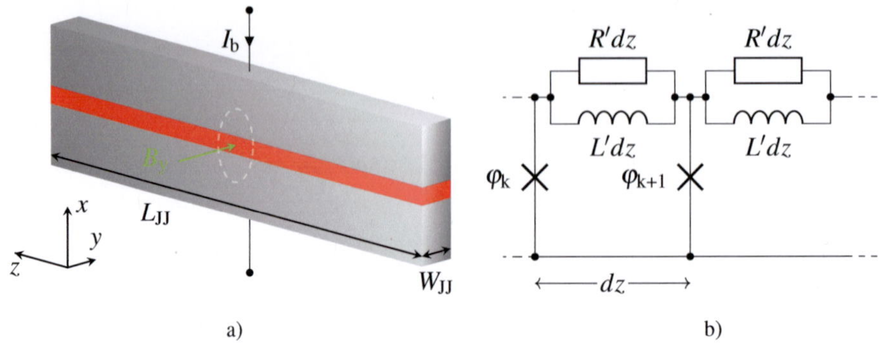

Figure 3.12: (a) Schematic illustration of a long Josephson junction with $L_{JJ} \gg \lambda_J$ and an applied magnetic field in plane with the barrier. (b) The equivalent circuit of this junction is the addition of many single short junctions.

is significantly larger than λ_L, but the width is still $W_{JJ} \leq \lambda_J$. This is called a long Josephson junction (lJJ). In order to describe the properties of such long junctions, the phase variation along the z-axis has to be derived, which can be done by separating the long junction into infinitely short portions as shown in [97, 98]. Following this derivation, the perturbed Sine-Gordon equation can be set up to

$$\frac{\Phi_0}{2\pi L^*}\frac{\partial^2\varphi}{\partial z^2} - \frac{\Phi_0 C_{\mathrm{tri}}}{2\pi}\frac{\partial^2\varphi}{\partial t^2} - j_c \sin\varphi = -j + \frac{\Phi_0}{2\pi\zeta}\frac{\partial\varphi}{\partial t} - \frac{\Phi_0}{2\pi\zeta_s}\frac{\partial^3\varphi}{\partial z^2 \partial t}, \tag{3.61}$$

where L^* describes the specific inductance, ζ is the specific resistance and ζ_s represents the specific surface resistance. Normalizing this equation with $z = \tilde{z}\lambda_L$ and $t = \tilde{t}\omega_{\mathrm{plasma}}^{-1}$ yields

$$\beta_{SG}\varphi_{\tilde{z}\tilde{z}\tilde{t}} + \varphi_{\tilde{z}\tilde{z}} - \varphi_{\tilde{t}\tilde{t}} - \alpha_{SG}\varphi_{\tilde{t}} = \sin(\varphi) - \frac{I}{I_c}. \tag{3.62}$$

In this equation, the subscripts indicate the partial differentiation with the constants α_{SG} and β_{SG} describing the damping due to the dissipation of quasi-particle tunneling and the surface resistance of the superconductors [97, 99]. Assuming small φ ($I = 0$), equation (3.62) can be linearized by substituting $\sin(\varphi) = \varphi$. An additional negligence of losses ($\alpha_{SG} = \beta_{SG} = 0$), simplifies equation (3.62) to the unperturbed linearized Sine-Gordon equation

$$\varphi_{\tilde{z}\tilde{z}} - \varphi_{\tilde{t}\tilde{t}} - \varphi = 0. \tag{3.63}$$

The solution of this equation can be found using a plane wave

$$\varphi(z,t) = \mathrm{e}^{\mathrm{j}(kz - \omega t)}. \tag{3.64}$$

Substituting this into equation (3.63) yields

$$\frac{\omega^2}{\omega_{\text{plasma}}^2} - \lambda_{\text{J}}^2 k^2 = 1 \tag{3.65}$$

which can be rewritten to

$$\omega = \sqrt{\omega_{\text{plasma}}^2 + \bar{c}^2 k^2}. \tag{3.66}$$

Here, $\bar{c} = \omega_{\text{plasma}} \lambda_{\text{J}}$ represents the so-called Swihart velocity, which is the maximum propagation velocity in the Josephson junction. Typically, $\bar{c}$ is in the range of $0.01..0.0001 \cdot c_0$ [81, 100] with c_0 being the speed of light in vacuum.

Obviously, equation (3.68) is a solution of the unperturbed linearized Sine-Gordon equation with a plane wave character. This can be interpreted as plasma waves traveling along the junction. Equation (3.66) gives the lower frequency limit of these plasma waves. Only waves with a frequency higher than ω_{plasma} can propagate inside the junction.

Relaxing the requirement of small φ values implies that the resulting equation is not linear any more. This leads to the unperturbed Sine-Gordon equation

$$\varphi_{\tilde{z}\tilde{z}} - \varphi_{\tilde{t}\tilde{t}} - \sin(\varphi) = 0. \tag{3.67}$$

A solution to this equation is given by

$$\varphi(\tilde{z}, \tilde{t}) = 4 \cdot \arctan\left(\mathrm{e}^{\pm \frac{\tilde{z} - \tilde{z_0} - \frac{v_{\text{z}}}{\bar{c}} \tilde{t}}{\sqrt{1 - (\frac{v_{\text{z}}}{\bar{c}})^2}}} \right). \tag{3.68}$$

This solution describes a time and space dependent phase change of 2π and thus represents the phase in presence of a fluxon (+) or antifluxon (-) traveling along a Josephson junction which behaves as a relativistic particle. Figure 3.13 depicts the behavior of the phase for $z_0 = 0$ and $v_{\text{z}} = 0$ at $t = 0$ with the blue line. The spatial extension of this resting fluxon is approximately $2\lambda_{\text{J}}$.

The fluxons can be accelerated by the bias current leading to a velocity that results from the equilibrium between the Lorentz force and the friction force

$$v = \frac{1}{\sqrt{1 + \left(\frac{4\alpha_{\text{SG}}}{\pi \cdot I / I_{\text{c}}} \right)}}. \tag{3.69}$$

The red line depicts the phase behavior for a fluxon with a velocity of $v_{\text{z}} = 0.9\bar{c}$.

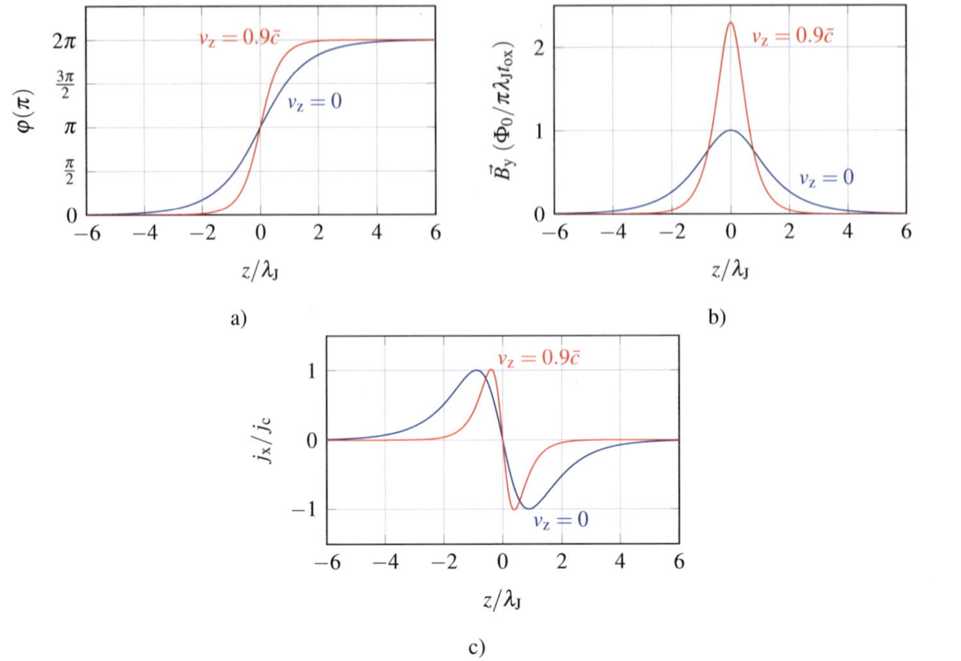

Figure 3.13: Phase variation along a long Josephson junction with the corresponding magnetic field and current density. The phase is depicted in (a) with a phase change of 2π, corresponding to a single flux quantum. The corresponding magnetic field is depicted in (b) and the current density can be seen in (c). Each of these graphs contains the calculations of these quantities for two fluxon velocities, the stationary case of $v_z = 0$ and $v_z = 0.9\bar{c}$.

Due to the high velocities of the fluxon close to the speed of light, they are subject to relativistic contraction. This contraction is also visible for the flux density B_y and the current density j_x normalized to the critical current density. These two quantities can be calculated from (3.68) using the relations (3.40) and Ampère's law $\frac{\partial B_y(z)}{\partial z} = -\mu_0\, j_x(z)$ and are also depicted in 3.13.

The two excitations shown above manifest themselves in the current-voltage characteristic of the long junctions. Exiting plasma waves in a long junction can be achieved by a magnetic field, spatially modulating the Josephson current in the long junction. Due to the finite length of the junction, the plasma wave is in parts reflected at the junction boundary, creating a standing wave if the damping is small enough ($\alpha_{SG} \cdot L_{JJ}/\lambda_J < 1$). This resonant behavior of a standing wave with the wavelength λ can be observed, if the condition

$$L_{JJ} = n\frac{\lambda}{2} = n\frac{\pi}{k} \tag{3.70}$$

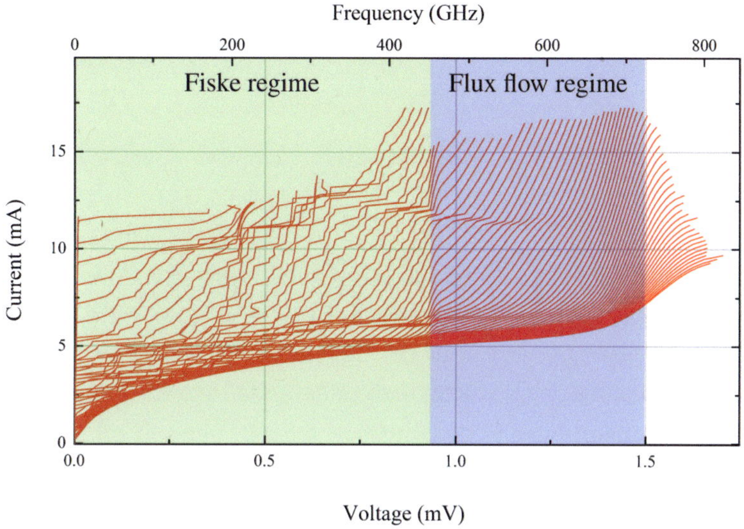

Figure 3.14: Measured set of current-voltage characteristics of a long Josephson junction. Each of these curves is recorded for a different magnetic field. The curves at low voltages correspond to small magnetic fields. Increased magnetic fields shift the curves towards higher voltages. The Fiske regime is indicated by the green and the flux-flow regime by the blue areas.

is satisfied, for a magnetic flux amount of $n \cdot \Phi_0/2$ present in the junction, where n is an integer. This resonance manifests itself as current steps in the current-voltage characteristic as can be seen in figure 3.14. These steps were first reported by Fiske [101] and are therefore called Fiske steps. The steps occur at the voltages

$$V_{\mathrm{n}} = \frac{\omega_{\mathrm{plasma}}}{2\pi} \frac{\Phi_0 \lambda_{\mathrm{J}}}{2L_{\mathrm{JJ}}} \cdot n. \tag{3.71}$$

Characteristic for these steps is that they can exhibit very small differential resistances for instance $R_{\mathrm{d,bias,FS}} = 0.002\,\Omega$ [102], which depend on the quality factor of the resonator.

The second kind of excitation is the excitation by fluxons. If a sufficiently large magnetic field is applied in plane with the barrier, fluxons can enter the junction from the edges [81]. If the Lorentz force is large enough to overcome the pinning forces, fluxons are accelerated in one direction towards one end of the junction depending on the polarity of the currents and magnetic field. Due to the finite length of the junction, the fluxons will eventually arrive at the opposite boundary, where they leave the junction emitting microwave radiation. Therefore, a continuous unidirectional flux-flow is present in the junction, which is why this type of device is called flux-flow oscillator (FFO). The voltage across a flux-flow oscillator is given by

$$V_{\mathrm{FFO}} = B_{\mathrm{y}} t_{\mathrm{ox,m}} v, \tag{3.72}$$

where the speed of the fluxons is noted by v [81]. An increased magnetic field hence leads to higher voltages of the flux-flow oscillator which corresponds to a higher frequency (3.26). For a given magnetic field, the motion of the fluxons manifests itself in the current-voltage characteristic as flux-flow steps. Such a device is thus a tunable microwave oscillator which can emit microwave radiation up to a frequency of $v_{\max} = \Phi_0 V_0$, where V_0 is the bias voltage. It was therefore proposed by Nagatsuma [98, 103] to employ Josephson oscillators as an on-chip integrated microwave source. Real devices have been shown with frequencies as high as 700 GHz with niobium based FFOs, limited by the gap voltage [104].

The characteristic quantities of an oscillator besides the oscillation frequency are the output power and the oscillation linewidth. The output power of the flux-flow oscillator connected to a load impedance R_{L} (see figure 3.15) can be calculated using

$$V_{\mathrm{L}} = V_{\mathrm{osc}} \frac{R_{\mathrm{L}}}{Z_{\mathrm{FFO}} + R_{\mathrm{L}}} \tag{3.73}$$

to

$$P_{\mathrm{L}} = \frac{V_{\mathrm{L}}^2}{2R_{\mathrm{L}}} \tag{3.74}$$

with the oscillator voltage V_{L}, and the oscillator output impedance Z_{FFO}. An intuitive image for the linewidth of the emitted radiation of the FFO can be derived when considering the fact that the output frequency of the flux-flow oscillator is defined by the Josephson equations. According to them, the frequency is determined by the bias voltage of the oscillator. Assuming an ideal bias voltage or current with zero fluctuations, the linewidth of the output radiation would only be determined by internal current fluctuations [105]. However, in a real application, the biasing is subject to variations itself. Assuming a current biasing scheme, a noisy current leads to a varying voltage due to the finite differential resistance of the current-voltage characteristic. From this derivation, it is obvious, that the linewidth of the flux-flow oscillator strongly depends on the differential resistance values

$$R_{\mathrm{d,bias}} = \frac{\partial V_{\mathrm{bias}}(I_{\mathrm{bias,CTRL}})}{\partial I_{\mathrm{bias}}} \text{ and} \tag{3.75}$$

$$R_{\mathrm{d,CTRL}} = \frac{\partial V_{\mathrm{bias}}(I_{\mathrm{bias,CTRL}})}{\partial I_{\mathrm{CTRL}}}. \tag{3.76}$$

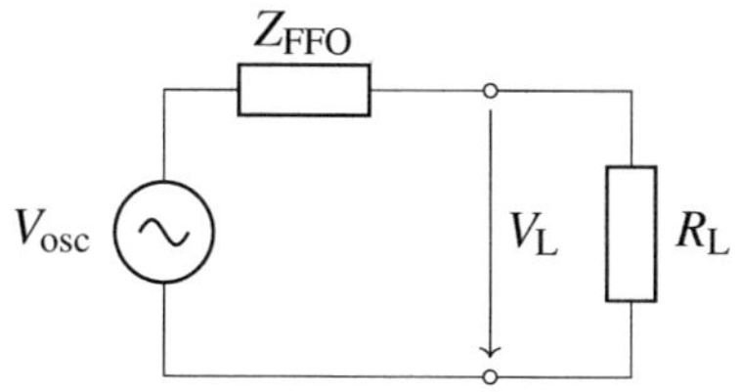

Figure 3.15: Equivalent circuit of a flux-flow oscillator connected to a load resistance. The FFO is modeled as a voltage source with an output impedance Z_{FFO}. It is connected to a load impedance R_L. This circuit needs to be considered for a maximized power transmission from the FFO to the load.

Both of these quantities can be obtained from the current-voltage characteristics of a flux-flow oscillator. The obvious $R_{d,bias}$ directly reads from the slope of the current-voltage characteristic as depicted in figure 3.14. The value of $R_{d,CTRL}$ needs to be measured separately. Both values determine the resulting FFO linewidth [11]. Obviously, the Fiske regime is beneficial for a narrow linewidth of the output radiation due to the high differential resistance. A bias current subjected to noise and fluctuations will only result in a negligible voltage and hence frequency variation.

However, the Fiske regime has two drawbacks. First, due to the step-like appearance, a continuous frequency tuning is not possible, but only discrete frequencies can be achieved. The second drawback concerns the upper frequency limitation. Since the Fiske steps can only be measured for voltages below a certain level, this regime is limited in operation frequency to values lower than ≈ 450 GHz.

Above the Fiske regime, the FFO is driven into the flux-flow regime, which enables a continuous adjustment of the frequency, with the drawback of a smaller differential resistance, hence a broader output spectrum. Typical linewidth values found in literature are between 10 and 100 MHz for operating the FFO on flux-flow steps and values between 0.4 and 10 MHz for an operation on Fiske steps [106, 107]. These free-running linewidths can furthermore be influenced by geometrical parameters [105]. It can be influenced and improved by e.g. the critical current density [108] or the junction width [109], however, a substantial decrease of the FFO linewidth is possible by the implementation of a phase-locked loop with the flux-flow oscillator. In this case, the linewidth significantly drops to values as low as 1 Hz [104] on Fiske steps.

The output power of a flux-flow oscillator is typically in the nW to μW-range [16, 98, 106]. In order to integrate a flux-flow oscillator in an RF-environment (typically microstrip transmission lines), and provide the mixer with sufficient pumping power, the output impedance needs to be determined in order to achieve matching with the adjacent transmission line network.

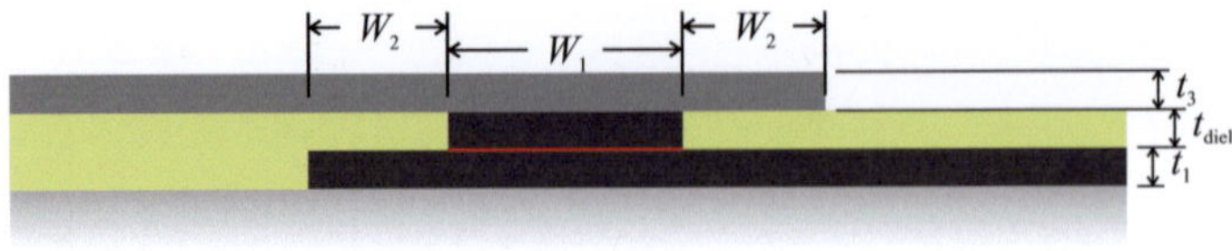

Figure 3.16: Schematic drawing of the output cross section of a flux-flow oscillator. The ground electrode is depicted in dark gray, the dielectric in green and the wiring layer is colored in lighter gray. The important quantities for the calculation of the output impedance of the FFO are given in the figure.

For the RF properties, the flux-flow oscillator is considered as a distributed transmission line and the output impedance can be approximated by the junction's characteristic impedance [103, 110]. According to equation (2.24), the only parameters that are required, are the specific inductance L' and capacitance C'. These parameters can be calculated for a given junction cross section [111]. Therefore, the junction cross section needs to be considered, which is depicted in figure 3.16. The values of interest can then be calculated by

$$\frac{1}{L'} = \frac{1}{\mu_0}\left(\frac{W_1}{h_1} + \frac{2W_2}{h_2}\right) \tag{3.77}$$

with

$$h_1 = \lambda_L \coth\left(\frac{t_1}{\lambda_L}\right) + \lambda_L \coth\left(\frac{t_2}{\lambda_L}\right) \tag{3.78}$$

and

$$h_2 = t_{\text{diel}} + h_1 . \tag{3.79}$$

The specific capacitance can be determined by

$$C' = C'_{\text{tri}} \cdot W_1 + \frac{2\varepsilon_0 \varepsilon_r W_2}{t_{\text{diel}}} . \tag{3.80}$$

The geometric parameters for W_i and d_i are given in figure 3.16 which depicts the cross section of the flux-flow oscillator. ε_r represents the dielectric constant of the applied dielectric. For this work silicon monoxide was used with $\varepsilon_r = 5$.

Obviously, the widths of the idle region W_2 and the active region W_1 contribute to the output impedance, where a larger idle region results in lower output impedances. Since C' is larger than L', the output impedance of a flux-flow oscillator is typically low which makes it hard to match it to an adjacent microstrip network. It is hence beneficial for the matching, if the width of the active region of the flux-flow oscillator is reduced along with a reduced width of the idle region. In order not to decrease the total area, the junction is tapered at the very end. This can be seen in figure 3.17, which schematically depicts a top

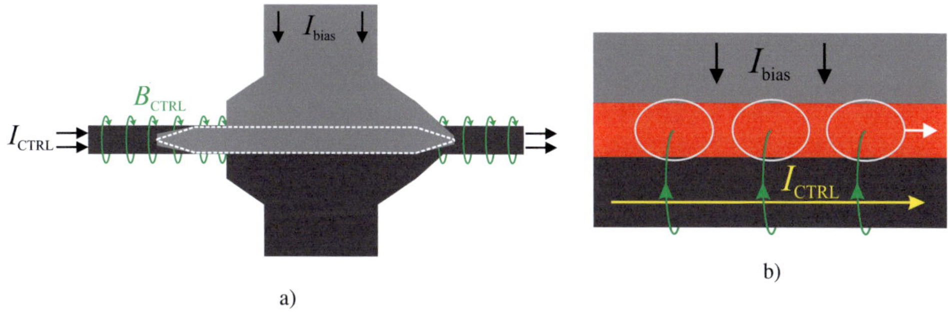

Figure 3.17: Schematic top view (a) and cross section (b) of a flux-flow oscillator. The wiring layer is depicted in lighter, the ground electrode in darker gray. The junction area is illustrated by the dashed white line. The FFO output is assumed to be at the right boundary of the junction. With an applied bias current as given in the figure, the fluxons generated by the magnetic field are accelerated towards the right boundary of the FFO.

view of a flux-flow oscillator and a cross section. This tapering is in this work realized as a linear taper, however, there are also other approaches to realize this, e.g. an exponential taper [112]. This tapering is furthermore responsible for the suppression of unwanted small resonant structures, which are superimposed to the Fiske steps [107, 113]. Figure 3.17 also depicts another feature which needs to be considered for the design of flux-flow oscillators: the unbiased tail. This feature is owed to the inhomogeneous current distribution in the long Josephson junction [103, 114]. Following this derivation of Nagatsuma et al., a current along the junction electrodes is induced by the bias current, which is accompanied by a magnetic field which is called the self-field. This self-field decreases the slope of the current-voltage characteristic of the FFO and thus result in a larger linewidth. This effect can be suppressed with a bias line geometry as indicated in figure 3.17, where the width of the bias line is smaller than the length of the oscillator. This leaves the tail of the oscillator unbiased and yields a smaller linewidth due to an increased differential resistance.

The generation of the required external magnetic field for the injection of fluxons in the junction can be done by an external coil. However, the mixer on chip is also susceptible to magnetic field, which is why the magnetic field for the FFO has to be created locally on chip. This is typically achieved by a control line, which passes a current along one of the electrodes of the FFO as indicated in figure 3.17. The magnetic field that is generated by this current is in close vicinity of the barrier and can thus generate the required fluxons in the junction.

Conclusion

The long Josephson junction is introduced in this section. This junction exceeds the Josephson penetration depth in one dimension $L_{JJ} \gg \lambda_J$. Therefore, the phase can vary along the length of the junction, which is described using the Sine-Gordon equations. These long junctions can be utilized as local oscillators. Both, the SIS mixer as well as the local oscillator can thus be realized in the same fabrication technology and can thus be easily integrated on a single chip, resulting in a superconducting integrated receiver. The local oscillator is operated in the flux-flow regime for higher frequencies, and is therefore called a flux-flow oscillator. The output impedance of such a flux-flow oscillator is discussed, since the FFO needs to be matched to the adjacent transmission line network. The realization of an integrated receiver comprising a flux-flow oscillator connected to a SIS mixer requires the fabrication technology to comply with particular demands that stem from the properties of these devices, which will be introduced in the following section.

3.3 Technological requirements for superconducting integrated receivers

The previous chapters have introduced the necessary basics for understanding the setup and operation of a superconducting integrated receiver. However, if such a device is to be realized, the fabrication technology has to fulfill a number of requirements in order to provide a quality that fits the particular demands of a superconducting receiver. Figure 3.18 depicts the schematic setup of a superconducting integrated receiver. As opposed to a mixer with an external local oscillator, the signal combination of the local oscillator signal with the received signal is realized on chip, which is why the transmission line network on chip needs to provide for the low loss signal combination on chip.

Achieving the low loss signal transmission and processing requires the energy gap of the superconductors to be large enough. The mixer can basically operate up to twice the gap frequency which is why the limiting element in this case is given by the flux-flow oscillator. A flux-flow oscillator can generate frequencies up to approximately $\nu_{gap} = 483.6\,\mathrm{GHz/mV}$. In case of a niobium based trilayer, this limit is 700 GHz. The usage of a superconductor with a higher energy gap can increase the maximum operation frequency up to e.g. 1.4 THz in the case of niobium nitride. Therefore, the fabrication process needs to allow for the deposition of films with high critical temperatures in order to maximize the energy gap (see equation (3.9)). This is however not only required for the SIS trilayer, but also for the wiring layer which is applied on top of the patterned trilayer, providing the necessary connections to the

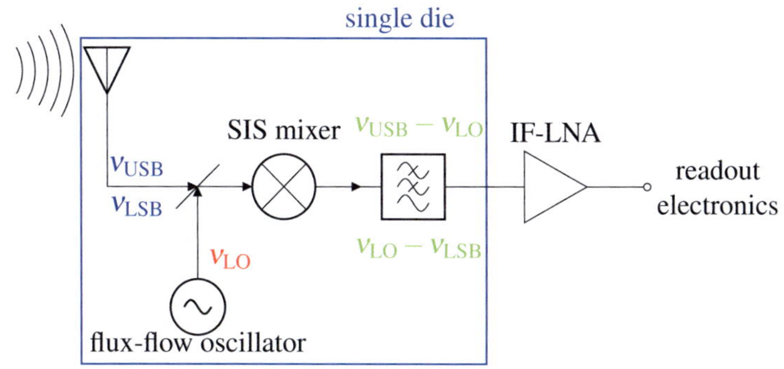

Figure 3.18: Schematic setup of an integrated receiver. The components included within the blue box are all realized on a single die. The signal is received by an antenna, the local oscillator signal is generated by the flux-flow oscillator. After the mixer, the low pass filtered signal is transferred from the die to the IF periphery.

particular components. The wiring layer also creates the microstrip and therefore needs to be able to transmit the signal with the lowest possible losses. Another source of losses in power transmission is the reflection due to impedance mismatch. In order to decrease the losses in the microstrip lines the used material should exhibit an energy gap that is high enough to carry the signal as well as a minimum surface impedance (see equations (3.16)-(3.20)). Furthermore, in order to achieve a reliable matching with low reflection, the deposition process needs to yield reproducible film qualities. The wiring layer is, as opposed to the trilyer electrode, not deposited on an atomically flat single crystal substrate, but on top of multiple materials, specifically on top of the electrodes as well as the dielectric, which in this case is silicon monoxide (SiO). This requirement poses a significant challenge to the deposition technology.

The quasiparticle leakage current in conjunction with the degree of the non-linearity of the current-voltage characteristic of the mixer dictates the noise temperature of the mixer [16]. The deposition of the trilayer is therefore required to yield an interface between the superconductors and the barrier that is as ideal as possible in order to avoid unwanted energy states in the vicinity of the barrier. This results in a high ratio of the sub-gap to normal state resistance and sharp current rise in the current-voltage characteristic at the gap voltage.

The patterning needs to allow for the realization of the required dimensions in the particular layers. In particular the junction definition and the wiring layer are the most crucial steps. The junctions need to be as small as possible in order to minimize the mixer capacitance. However, care has to be taken since the normal state resistance for a given j_c increases for smaller junction areas. For the realization of the microstrip lines, two parameters need to be controlled in order to be able to adjust a characteristic impedance of the resulting microstrip in a reproducible and constant manner: the height of the dielectric and the width of the microstrip (see equation (2.29)). High characteristic impedances are achieved by either large thickness of the dielectric or a narrow microstrip. The thickness of the dielectric thus needs to be controlled and furthermore be selected such that the wiring layer can overcome the edges of the dielectric at the points of contact but also provide sufficient isolation to the lower electrode and also account for the achievable characteristic impedances. The SiO thickness thus has to be selected specifically for the given process in particular the achievable electrode and wiring layer thicknesses.

The goal of this work is to develop a superconducting integrated receiver based on niobium nitride technology. This technology should enable the coverage of higher frequencies as compared to those based on pure niobium. This includes the development of a niobium nitride technology with an energy gap that is as high as possible, a critical current density as high as possible while maintaining a minimum leakage. Furthermore, the material parameters need to be extracted and provided to a simulation software, which enables the efficient design of an RF-circuit for the on-chip distribution of the THz signals.

4 Technology for high-quality NbN/AlN/NbN devices

The fabrication technology needs to meet the requirements that stem from the particular demands of a superconducting integrated receiver as presented in the previous chapter. Many of these requirements coincide with demands for other high quality devices, such as the maximization of the critical current density, which is also required in rapid single flux quantum (RSFQ) circuits. An increase in gap voltage can also be beneficial for superconducting quantum interference devices (SQUIDs) in order to decrease voltage noise due to temperature fluctuations. However, some requirements are different, especially the demand for a high ratio of sub-gap to normal state resistance. SQUIDs or voltage standards for example require highly damped junctions. Due to the various requirements that are posed by different devices, this chapter will solely refer to the quality values that are required by integrated receivers. The current state-of-the-art will be presented showing the properties of both, the most common technology the Nb/AlO_x/Nb technology and trilayers with electrodes exhibiting higher energy gaps. The state-of-the-art in fabrication technology will also be presented mostly regarding miniaturization of the junctions. The chapter will furthermore discuss the development of the patterning process of the junctions and the wiring layer and demonstrate the capabilities thereof with the Nb-based trilayer technology. Additionally, the NbN-based trilayer technology development will be shown and the patterning process will be evaluated considering the altered conditions, such as the increased layer thickness.

4.1 State-of-the-art

The mature niobium-based technology is nowadays frequently demonstrated with its theoretically maximum energy gap of 2.9 meV [115–120]. The critical current density for highly hysteretic junctions with low leakage was demonstrated frequently up to $14\,kA/cm^2$ [121, 122]. However, exceeding $15\,kA/cm^2$ already poses a significant challenge when low leakage currents need to be maintained. The upper limit of AlO_x barriers in the niobium technology seems to be around $20\,kA/cm^2$. However, introducing the aluminum nitride as barrier material has shown to enable a significant increase in critical current density without increasing the leakage compared to AlO_x barriers up to 30 kA/cm^2 [96, 123].

The miniaturization of the junction area is beneficial for many applications, including SQUIDs. Therefore, a big effort was made in order to achieve the smallest junction area possible. At this point, it has to be differentiated between the different fabrication process types: the classical window type junction which is essentially based on the selective niobium anodization process (SNAP) [124] and is the most flexible process concerning the design limitations. This process is however limited in junction miniaturization due to the fact that the junction-definition lithography involves the realization of resist stencils which define the size of the junction. Due to the lithographic process, this can be a challenge. The junctions with smallest areas have been realized using the cross-type junction process. Using this fabrication, junction areas as small as 0.36 μm^2 have been demonstrated using only photolithography [125]. This process is however limited in design, since it requires two stripes to cross each other, which is not beneficial for the integration of a small junction with a narrow microstrip line. In the context of this work, such strong geometrical restrictions do not seem acceptable, which is why the window type junction is used with one of the main goals being the miniaturization of the junction area.

One of the most important aspects of this work is the quality of the niobium nitride films and the trilayers that could be realized with them. In literature, the critical temperatures of NbN films vary strongly with a major influence stemming from the employed substrate and the substrate temperature during deposition. The following examples only refer to film thicknesses exceeding 200 nm. For decreased film thicknesses, the critical temperature drops to significantly smaller values [126]. The substrate resulting in the highest NbN film-quality is magnesium oxide (MgO) which yields critical temperatures of the NbN films exceeding 16 K [127, 128]. This is very close to the bulk value 17 K [129]. Other substrates mostly yield critical temperatures which are significantly lower than 17 K [117, 126, 130]. This manifests itself in a variation in the gap voltage of the Josephson junctions accordingly. Those junctions realized on MgO exhibit gap voltages as high as 6 mV with critical current densities as high as 70 kA/cm^2 [131]. This is however limited to the case of MgO as a barrier material. As previously described, AlN as a barrier material is favorable, due to the increased critical current density. The junctions realized on MgO with an aluminum nitride barrier typically exhibit gap voltages of only 5.2 mV [132, 133], however, in this case, the critical current density can be increased up to 100 kA/cm^2.

The energy gap of NbN films on other substrates are in most cases far below the values shown above. It was shown on silicon substrates that, using an elaborate pre-treatment of the substrate, the gap voltage of the NbN films could be increased up to 4.9 mV. However, till now, the critical current densities shown, are extremely low, excluding these junctions from a practical relevance for mixers so far [134]. NbN-based junctions have been reported on other types of substrates as well, such as silicon dioxide or fused silica, where the deposition process yielded only poor results concerning the gap voltage, which could not exceed values of $\approx 3.3\,\text{mV}$ [130]. On sapphire substrates, remarkable results have been obtained [135] with gap voltages as high $6\,\text{mV}$. This was however only possible by the incorporation of a MgO buffer layer. Nonetheless, this result is encouraging to continue the NbN deposition on sapphire substrates.

In order to reduce the complexity of the layer setup, this work aims at the realization of high quality NbN films directly on R-plane sapphire substrates without any pre-treatment or buffer layer. These NbN films are optimized according to the demands of superconducting integrated receivers, concerning the gap voltage, the leakage current and the critical current density. The energy gap and the critical current density need to be maximized, whereas the leakage needs to stay at low levels.

4.2 Junction and circuit fabrication

The realization of high-quality NbN-based Josephson junctions not only requires the SIS trilayer to be as perfect as possible, but also the patterning must not deteriorate the junction properties, enabling the measurement of the real trilayer quality. In particular, the wiring layer has to be separated from the bottom electrode, the layer thicknesses can not exceed certain levels in order to be able to use high-resolution electron beam lithography and the reproducibility needs to be ensured. Therefore, this section deals with the patterning process sequence, which is shown for niobium-based trilayers for which this process was originally developed [117]. This is done, since the niobium trilayer technology has matured to a level, where high-quality SIS trilayers can be realized reproducibly [115–117]. Therefore, the eligibility of the patterning can be qualified using Nb/Al-AlO_x/Nb trilayers.

4.2.1 The self-planarized process

The patterning process begins using a previously deposited SIS trilayer. The main feature of the self-planarized process is the ability to keep the layer thickness of the subsequently deposited layers at the minimum possible level. This is achieved by a consequent fabrication sub-sequence of etching and filling and will be explained in the following section using the cross-section depicted in figure 4.1.

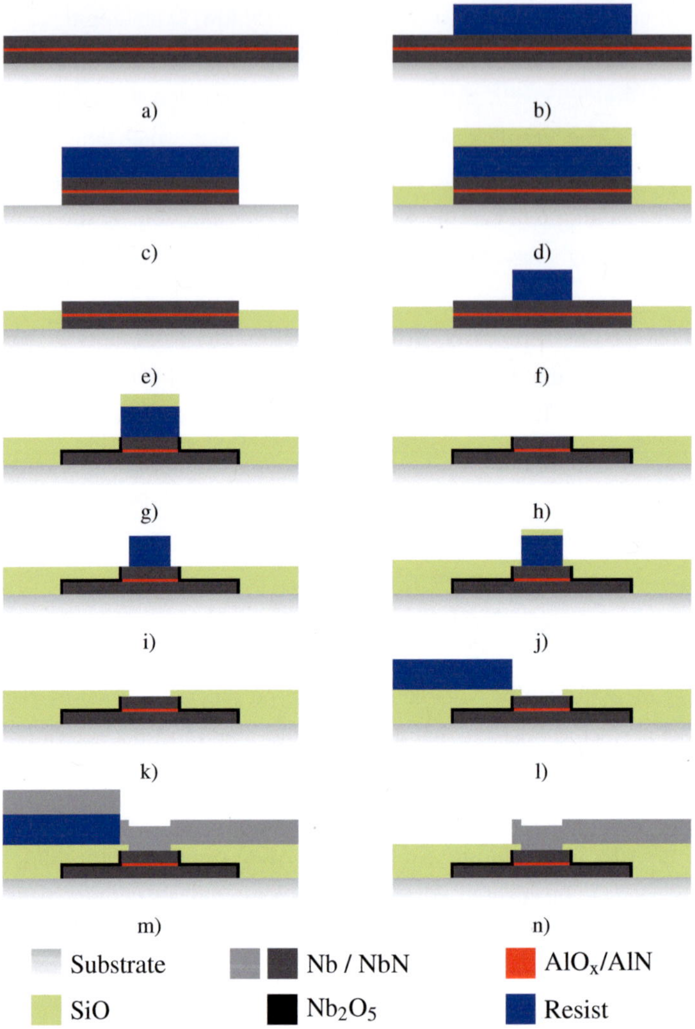

Figure 4.1: Fabrication sequence of the self-planarized process. Starting from a superconductor-insulator-superconductor trilayer on a substrate, first, the ground electrode is defined, followed by the junction definition and the wiring layer.

Starting from the trilayer on the substrate (figure 4.1a), the definition of the lower electrode is done first (figure 4.1b). The lithography defines the lateral dimensions of the lower electrode, and the subsequent etching patterns the entire trilayer accordingly. The niobium electrodes are hereby etched using reactive ion etching, whereas the isolating barrier is etched by ion milling. The resulting structure is depicted in figure 4.1c. Instead of stripping the resist immediately, a first layer of silicon monoxide is evaporated (figure 4.1d). The purpose of this layer is the planarization of the resulting steps at the lower electrode edges. Therefore, the thickness of this planarization layer is selected according to the thickness of the lower electrode $t_{\mathrm{Nb,bot}} = 90$ nm.

The silicon monoxide is then lifted off (figure 4.1e) and after another lithography for the junction definition (figure 4.1f), the counter electrode is etched, again using reactive ion etching, defining the lateral dimensions of the junction (figure 4.1g). After the etching, the whole die is immersed in a solution of ammonium pentaborate and ethylene glycol for anodic oxidation. Again, without stripping the resist, a SiO layer is deposited. In conjunction with the oxidized surface of the lower electrode, this SiO needs to provide a reliable electrical isolation between the bottom electrode and the wiring layer, preventing short circuits parallel to the junction. The isolation across the edge of the bottom electrode proved to be particularly critical in other processes, which is why the thickness of the SiO isolation layer needed to significantly exceed the thickness of the bottom electrode. However, due to the planarization layer, the isolation layer does not face steps underneath that needed coverage. Therefore, the thickness of the isolation layer is selected equal to the thickness of the counter electrode $t_{\mathrm{Nb,top}} = 90$ nm.

After lifting off the isolation layer, the structure is fully planar (figure 4.1h), which is beneficial for the following layers. In the next step, the lithography for a second isolation layer is done (figure 4.1i). This thin second isolation layer, which has a thickness of approximately 20 nm, is then deposited and structured by lift-off (figures 4.1j and 4.1k). The purpose of this layer is to ensure the isolation between the bottom electrode and the wiring layer. The necessity of this will be demonstrated later. Furthermore, the second isolation layer adds a degree of freedom when e.g. process requirements prevent the total thickness of the SiO to be selected for the ideal planarization. This means, that if a total SiO thickness was required, that exceeds the counter electrode significantly, the resulting step height can be reduced by 50 % as compared to the realization of the SiO in a single deposition process.

After the second isolation layer, the wiring layer can be applied. Since the wiring layer is the last deposited layer, it can be structured using lift-off (figure 4.1l, 4.1m and 4.1n), or by etching.

The minimum required thickness of the wiring layer is dictated by the maximum step height $\Delta_{\mathrm{step,max}}$ of the underlying structures. The thickness of the wiring layer needs to be at least the London penetration depth in order to provide a good superconducting connection. Even and especially across underlying steps, the effective thickness is however reduced, which is why, the wiring layer has to be deposited with a thickness that equals at least $\lambda_{\mathrm{L}} + \Delta_{\mathrm{step,max}}$.

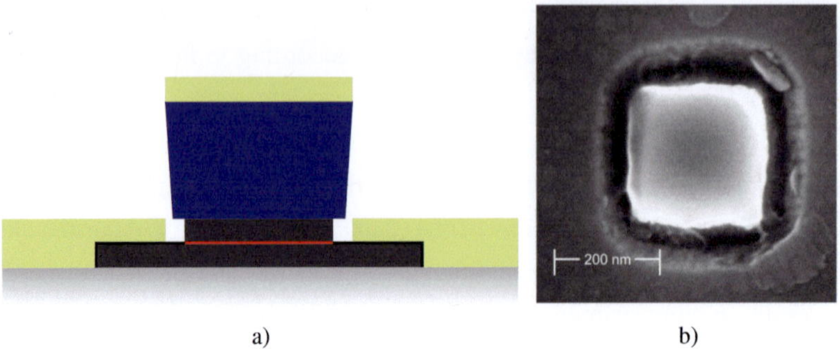

Figure 4.2: (a) Schematic cross section of a junction after the deposition of the first isolation layer, illustrating the isolation gaps surrounding the junction. (b) A SEM micrograph after the lift-off of the first isolation layer. The counter electrode surface is visible in the center of the image surrounded by a gap in isolation.

This is the particular benefit of the self-planarized process, since it allows for minimized step heights. Therefore, all layers, including the wiring layer, can be defined using high-resolution electron-beam lithography.

The consequent process sequence of etching and subsequent SiO-filling in the same lithography step allows the nearly completely planarization of the structures at intermediate fabrication steps without the use of chemical mechanical polishing. However, the etching of the electrodes is done using reactive ion etching, whereas the silicon monoxide is evaporated. Due to the isotropic nature of the reactive ion etching, the resist stencils that are the junction definition, are underetched. This is depicted schematically in figure 4.2a. The subsequent silicon monoxide evaporation is a very anisotropic process, which is why the underetched areas are not filled up completely by the SiO. This opens up gaps right around the junction, which can lead to short circuits, if the wiring layer is deposited directly on top. Figure 4.2b shows a scanning-electron micrograph (SEM) top view of the structure depicted schematically in figure 4.2a. The junction in the center is surrounded by the trench, before the SiO layer starts. This is why the second isolation layer is crucial to ensure a good electrical isolation between the bottom electrode and the wiring layer.

The result of the fabricated junctions is depicted as a cross-section in figure 4.3. This figure shows the SEM image of a cross-section of a realized junction. In the bottom part of the image, the substrate is visible, on top of which the bottom electrode is deposited. The edges of the electrodes exhibit the expected conical shape, which is due to the isotropy of the etching. The counter electrode on top has smaller lateral dimensions and also here, the conical shape of the edges can be recognized. Both electrodes are surrounded by SiO, which can not be identified optically in this image. However, on top of the counter electrode, the

Figure 4.3: Cross section of a junction after fabrication. On top of the substrate, the bottom electrode can be recognized as the widest bright structure. The subsequent narrower layer is the counter electrode followed by the wiring layer. The close-up in the right part of this figure depicts the right edge of the junction highlighting the consequence of the isolation gap of the first isolation layer.

window resulting from the second isolation layer becomes apparent. The wiring layer at the very top connects through this window and in the zoomed part of the image, a small kink in at the lower edge of the wiring layer is visible. This is due to the trenches that were discussed above, which are now filled with the second isolation layer and therefore, the wiring layer does not reach the bottom electrode.

Lithography One of the key aspects in the fabrication process is the lithography. It is crucial for the definition of the various geometries and the resist is furthermore subject to the subsequent fabrication processes. This becomes particularly important, if long etching procedures are necessary, or even a long etching procedure consisting of different etching techniques. In almost all cases, the resist has to withstand an etching procedure and allow for a consecutive lift-off process. Therefore, the resist must neither become too hard, nor be completely removed during e.g. an etching procedure. Therefore, the selection of the resist for the particular fabrication step is of utmost importance.

The process as described above can be done using either, photo- or electron-beam lithography. The resolution limit of the photolithography system is at approximately 1 μm. Therefore, sub-micrometer structures have to be defined using electron-beam lithography. Test structures with lateral dimensions above 1 μm that were realized in order to characterize the trilayers were done only using photolithography, whereas the receiver devices were fabricated using electron-beam lithography exclusively, due to the ultimately high level of design flexibility.

This fabrication process exhibits an acceptable turn-around time and obeys a design accuracy that allows the realization of square shaped $250 \times 250\,\text{nm}^2$ sized junctions as shown in figure 4.2b. Therefore, the results of the optical investigation of the structures at intermediate fabrication steps are satisfying and encourage the expectation of high-quality Josephson junctions.

4.2.2 Evaluation of Nb-based trilayer quality

After a successful fabrication, the realized junctions need to be measured electrically in order to determine their quality. Therefore, the junctions need to be cooled to their operation temperature. All measurements shown here are performed at liquid helium temperature ($T = 4.2\ \mathrm{K}$).

A measurement setup was developed [116, 117] and optimized which ensures a reliable, low noise measurement of the junctions. The schematic setup of the measurement system is depicted in figure 4.4. The device under test (DUT) is mounted on a dipstick, which is then inserted into a liquid helium dewar. Controlled by a computer, the digital to analog converters (ADCs) control a current source, which is connected to the DUT through RC-filters. The identical filters are used to connect the voltage amplifiers, which are read-out by analog to digital converter (ADCs). The DACs and ADCs have a precision of 16 bit with a sampling rate of 333 $\mathrm{kS/s}$, which is sufficient for the devices measured here.

In order to provide a magnetic field to the DUT, a copper coil can also be mounted on the dipstick, surrounding the DUT completely. This coil has a number of 9195 windings and can create magnetic fields as high as 0.162 $\mathrm{mT/mA}$. In order to set the magnetic field, the coil current is also computer controlled. For almost all measurements, the software "GoldExi" written by Dr. E. Goldobin [136] was employed. In order to screen external, unwanted magnetic fields, the entire lower end of the dipstick, where the DUT is mounted, is surrounded by a "cryoperm" cylinder.

The setup provides 6 identical and independent channels that enable the measurement of 6 single junction devices during one thermal cycle. Figure 4.5a depicts the current-voltage

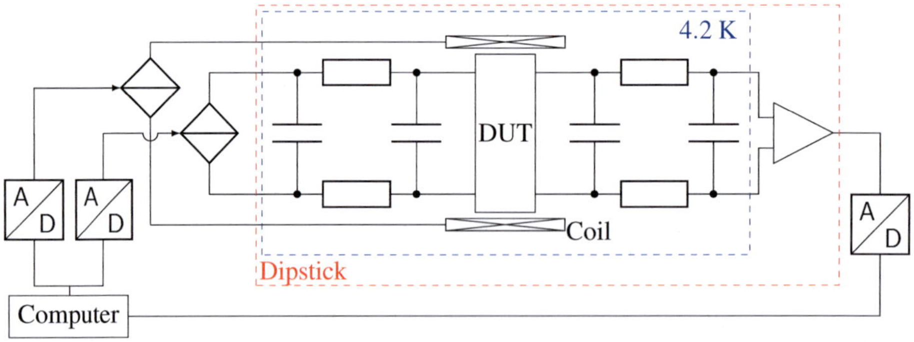

Figure 4.4: Circuit of the dipstick measurement setup. The computer controlled current sources are connected through cold filters to the device under test (DUT). The voltage is read by instrumentation amplifiers, and fed to the computer. The magnetic coil providing the magnetic field is also included in the dipstick, surrounding the entire DUT.

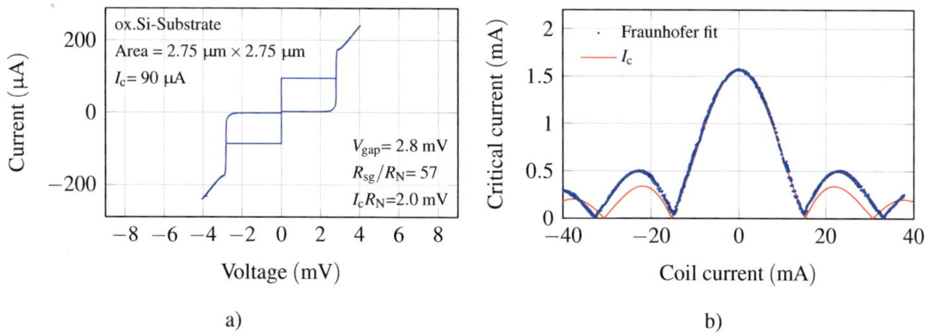

Figure 4.5: Measured current-voltage characteristic of a niobium based Josephson junction on an oxidized silicon substrate with the respective quality parameters in the inset. (b) The magnetic modulation of the Josephson critical current along with a Fraunhofer fit.

characteristic of a niobium-based Josephson junction, fabricated with the process described above and measured at 4.2 K. The gap voltage is 2.8 mV, which is close to the optimum, where the small reduction is most likely due to slightly thinner electrodes. The current rise at the gap voltage is almost vertical, with a negligible gap smear, which proves a nearly perfect interface between the electrodes and the barrier. This is also supported by the ratio of the sub-gap to normal state resistance ($R_{sg}/R_N = 44$), which proves a high quality of both, the trilayer itself, as well as the fabrication process, resulting in a very low leakage current of the $2.75 \times 2.75\ \mu m^2$ sized junction at an intermediate critical current density of $1.2\ kA/cm^2$. The theoretically ideal value of the I_cR_N-product for niobium electrodes is 2.3 mV. The measured $I_cR_N = 2$ mV is close to the ideal value, indicating nearly perfect tunneling properties. Furthermore, the measured critical current depending on the magnetic field is depicted in figure 4.5b along with a Fraunhofer fit. The suppression of the critical current is clearly visible, the fit to the Fraunhofer spectrum however does deviate from the measured values. This behavior however cannot be explained by a misaligned junction to the magnetic field orientation, the exact reason for this deviation is still unclear.

Determining the reproducibility of the fabrication process can be done by measuring multiple junctions. The fabrication variation is hereby expected to manifest itself in variations in the critical current due to inaccuracies during junction definition. Assuming a constant critical current density across the chip, variances in the junction area will lead to variances in the critical current. Therefore, a characterization chip was fabricated, containing a set of 30 junctions with different areas ranging from 1 μm^2 up to 71 μm^2. All of these junctions have been measured and the results of these measurements is shown in figure 4.6a.

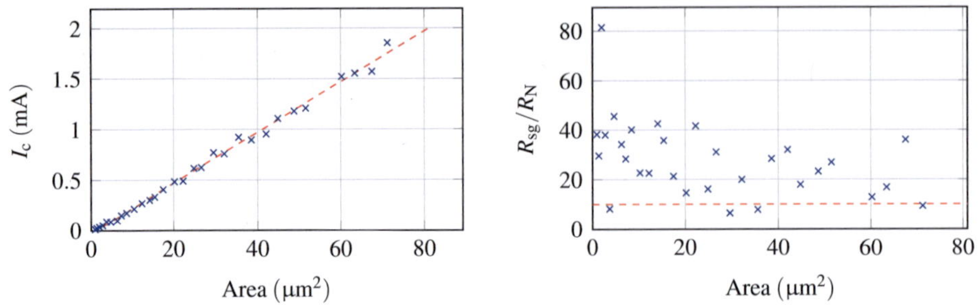

a) b)

Figure 4.6: Evaluation of I_c and the R_{sg}/R_N ratio for many short junctions. (a) The critical current clearly follows a linear fit which is indicated by the red dashed line. (b) The R_{sg}/R_N ratio exhibits large values and for almost all junctions is above the quality criterion of 10 (horizontal red dashed line).

As expected from theory, the critical current of the junctions follow a linear dependency on the junction area. Furthermore, almost all of the measured junctions showed an excellent quality regarding the leakage as depicted in figure 4.6b. The data presented here allows the conclusion of the self-planarized process being a fabrication process that enables a high quality realization of Josephson junctions with a high yield and reliability for micrometer sizes.

For miniaturization purposes, electron-beam lithography is applied. The result of a circular shaped junction can be seen in figure 4.7. The measured current-voltage characteristic proves the high quality. The gap voltage is slightly reduced as compared to the previously shown junctions. Also, a slight increase in the gap smear is visible and the critical current is suppressed. All these decreases in quality may be due to the fact that the junction radius in this case was only 100 nm, where the lateral dimensions already become comparable to $\lambda_{J,Nb} = 90$ nm. However, the junction shows an uncorrected $j_c = I_c/A = 8\,kA/cm^2$ and still exhibits an excellent ratio of the sub-gap to normal state resistance of 19. Figure 4.7b depicts the SEM top view of that junction after the wiring layer has been deposited already. Clearly visible are the niobium grains and in the center, a circular shade is visible. This shade is the projection of the junction edge through the wiring layer. The wiring layer is deposited onto very small steps underneath, which is why the top of the wiring layer does not exhibit distinct, sharp features, making it hard to recognize these steps. The junctions can thus be realized with lateral dimensions down to ≈ 200 nm.

However, the benefit of the planarization is not limited to the junction miniaturization, but includes the applicability of high-resolution electron-beam lithography in every layer. This is demonstrated by the miniaturization of structures in the top layer, the wiring layer.

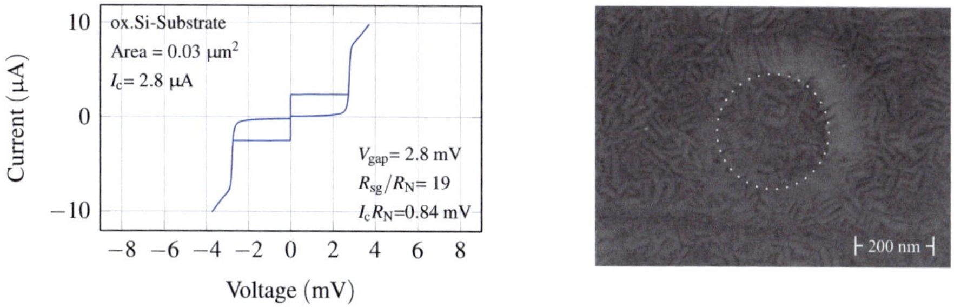

a) b)

Figure 4.7: (a) Current-voltage characteristic of a circular Josephson junction with a radius of $r = 100\,\text{nm}$. A top view of the wiring layer on this junction is depicted in (b).

Certain Josephson junction devices [137] require feature sizes in the wiring layer with lateral dimensions down to the nm-range. An example thereof is depicted in figure 4.8. In this example, a long Josephson junction is fabricated which is connected to the bias through a set of narrow stripes that all connect on a common resistor. This biasing setup aims at a bias current homogenization [117]. Furthermore, from the opposite side of the junction, so-called current injectors are realized. These current injectors are to create not only integer Josephson fluxons but fluxons that carry a fraction of the flux quantum [138]. According to theory, the ideal current injector exhibits a width of zero [137], which is why the injectors are to be realized as narrow as possible. Assuring that the injector current is flowing through the counter electrode instead of the wiring layer, the bias stripes and the current injectors must not be connected in the wiring layer, but only through the counter electrode. The realization thereof can be seen in the close-up in figure 4.8. The current injectors as well as the bias stripes were realized with a width of only 500 nm, the gaps between the current injectors and bias stripes is as narrow as 200 nm. Although, the evaluation of these narrow structures proves to be challenging, the miniaturization capabilities of the fabrication process are demonstrated. Micrometer sized current injectors were successfully employed to create fractional vortices using this fabrication technology [MMS⁺18].

Conclusion

In this section, the fabrication process is demonstrated. The planarization properties of the process in conjunction with the two-step isolation results in minimum film thicknesses of the consecutive layers. This enables the application of electron-beam lithography in every single layer, thus the miniaturization of structures is not limited to particular layers. The capability of the process was demonstrated using conventional niobium-based trilayers.

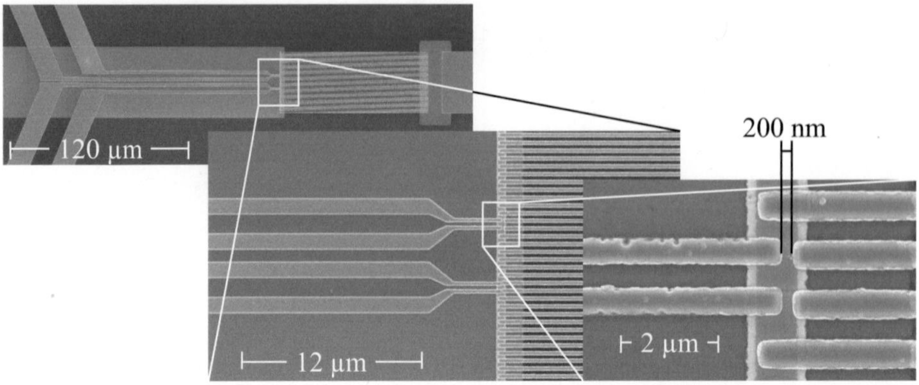

Figure 4.8: SEM images of a long junction with current injectors. The long junction is located in the center of the top left image. In the close-ups towards the right bottom end, the junction center is shown with a striped current feed coming from the right of the junction and 500 nm wide current injectors attached to the junction from the left. A gap of 200 nm is realized between the current injectors from the biasing fingers in the wiring layer.

The process results in a high yield with excellent quality parameters for micrometer sized Josephson junctions. The application of electron-beam lithography allows for the miniaturization of the junctions down to diameters of 200 nm, maintaining the high quality of the trilayers. The applicability of electron-beam lithography in all layers was proven by the realization of sub-micrometer features in the wiring layer with dimensions of 500 nm stripe width and 200 nm gap width. The fabrication process thus proved its suitability for the application in realizing high-quality devices.

4.3 NbN technology

The patterning process has proven to be suitable for the fabrication of miniaturized high-quality Nb-based SIS devices. However, as demonstrated in section 3, the frequency of these devices is limited due to the energy gap in niobium. In order to increase the operation frequency, superconductors with high energy gaps need to be employed. This section demonstrates the optimization of the niobium nitride technology with respect to the particular needs of a superconducting integrated receiver.

4.3.1 Deposition of NbN and AlN thin films

For an integrated receiver with optimum performance, the whole trilayer needs to exhibit excellent quality. This is not limited to the energy gap of the electrodes, but also includes the interface with the barrier. This interface needs to be very close to the ideal step-like structure in the energy diagram, in order to yield good performance. However, the optimization of the trilayer as a whole is very time-consuming, which is why the properties of the single NbN layers are optimized separately.

Deposition process

In order to achieve high quality interfaces, the deposition of the whole trilayer has to be done in-situ. Therefore, a deposition system has been developed, built [117] and optimized, providing the necessary conditions to achieve a maximized quality. A sketch of the deposition system is depicted in figure 4.9. It comprises three vacuum chambers, all separated by gate valves. The first chamber is a load lock, allowing for the loading and unloading of the samples without breaking the vacuum in the recipients. The load lock is connected to the first recipient, which has a niobium target installed. A movable heater stage, that is capable of reaching a temperature of more than 900°C, is installed. The heater stage can move the substrates to the second recipient, which has an aluminum target installed. This movable heater stage allows the in-situ deposition at elevated temperatures without the need for heating and cooling in between deposition processes.

Each chamber is equipped with a set of turbomolecular pump and a forepump, achieving base pressures down to $2 \cdot 10^{-7}$ mbar. The metal targets are mounted to a magnetron, which is driven by a 1 kW DC sputtering source. The metal-nitrogen compounds are reactively sputtered in an argon (Ar) nitrogen (N_2) atmosphere.

In order to perform a deposition, the process gases are introduced, where the gas flow is controlled by the means of needle valves. The pumping cross-section can be throttled using a butterfly valve right before the turbomolecular pump. The total gas pressure in the chamber is given by

$$p_{\text{total}} = p_{\text{par,Ar}} + p_{\text{par,N2}}, \tag{4.1}$$

where $p_{\text{par,i}}$ denote the partial pressures of argon and nitrogen respectively. The sputtering process is started by igniting a plasma in the chamber and the process pressure establishes after a short settling time

$$p_{\text{process}} = p_{\text{par,Ar}} + p_{\text{par,N}_2} - p_{\Delta\text{N}_2} = p_{\text{total}} - p_{\Delta\text{N}_2}. \tag{4.2}$$

The term $p_{\Delta N_2}$ refers to the nitrogen which reacts with the sputtered niobium to the desired niobium nitride. Since it is bound to the solid material from the gas phase, this value is called nitrogen consumption.

Figure 4.9: Three-dimensional sketch of the NbN/AlN DC magnetron sputtering system. In the top left part of the image, the load lock is located, attached to the NbN chamber in the center of the image. In the NbN chamber, a movable heater stage is installed, on which the samples are dropped from the load lock transfer rod. The heater can be moved between the AlN and NbN chamber while maintaining its temperature. The three chambers are separated from each other and at each chamber, a pumping combination of a forepump and turbomolecular pump is attached.

Depending on the deposition current, the temperature and the argon and nitrogen pressure, this value is very sensitive to the deposition conditions. This already illustrates the complexity of the reactive deposition process.

Figure 4.10 depicts the resulting current-voltage characteristics of the plasma discharge in the recipient with the niobium target for an argon pressure of $p_{\mathrm{Ar}} = 1.4 \cdot 10^{-3}$ mbar and a nitrogen pressure of $p_{\mathrm{N_2}} = 0.47 \cdot 10^{-3}$ mbar. Obviously, the reactive curve (blue line in figure 4.10) starts from a high voltage regime, then exhibits a section with a negative differential resistance right before it approaches the curve with pure argon (red line). The part of negative differential resistance is shifted to higher currents with increasing nitrogen content. The high voltage regime results in NbN films in the so called nitrogen phase. The films fabricated in this phase exhibit a negative temperature coefficient of the resistance. The section, where the reactive curve approaches the argon curve is called the niobium phase, where the deposited film exhibits a lack of nitrogen and the resistance behavior of a metal. In between these two sections, the niobium nitride films are expected to exhibit the best properties concerning the critical temperature.

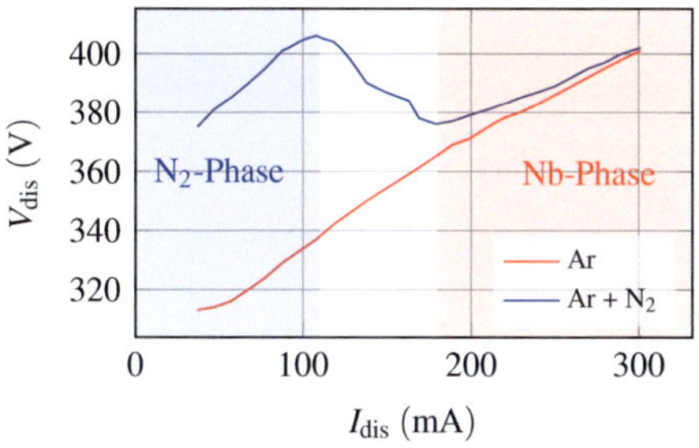

Figure 4.10: Discharge characteristic of the NbN sputtering system for the operation point for sapphire substrates($p_{Ar} = 1.4 \cdot 10^{-3}$ mbar, $p_{N2} = 0.47 \cdot 10^{-3}$ mbar). In the low current region, the resulting NbN films are in the nitrogen phase (indicated in blue) whereas in the high current regime, the resulting films are in the niobium phase.

From the large set of deposition parameters, the optimum value for each quantity has to be found in order to achieve the best performance of the resulting niobium nitride layers. The insulating barrier is created from aluminum nitride. The aluminum nitride is therefore deposited far in the nitrogen phase, in order to ensure a high electrical resistivity. The only requirement to the AlN layer is the electrically uniform isolation between the electrodes. However, since this barrier is only approximately $1-2$ nm thick, this isolation cannot simply be guaranteed by the optimization of the properties of the AlN. The underlying niobium nitride typically dictates the leakage through the barrier due to its surface roughness. Since the AlN sputtering is not a perfectly conformal process, a surface roughness of the NbN layer will result in leakage currents.

Therefore, the requirements to the niobium nitride layer do not only concern its electrical properties, but also the surface roughness. The optimization thereof will be discussed in the following where it is split between the optimization of the critical temperature and the surface roughness. Usually, these two parameters are anti-correlated. However, it will be shown that by the means of the nitrogen consumption, the critical temperature and the surface roughness can be optimized separately.

T_c optimization

Achieving high quality niobium nitride films has been subject to research for decades. Due to the almost perfect lattice match, the ideal substrate for NbN films ($a_{NbN} = 0.439$ nm) is MgO ($a_{MgO} = 0.42$ nm). The significant lattice mismatch of NbN and other substrates such as sapphire ($a_{sapphire} = 0.476$ nm) or silicon ($a_{Si\,[111]} = 0.384$ nm) substantially increase the requirements to the deposition technology. The NbN film quality on these substrates can be enhanced by the introduction of an aluminum nitride buffer layer.

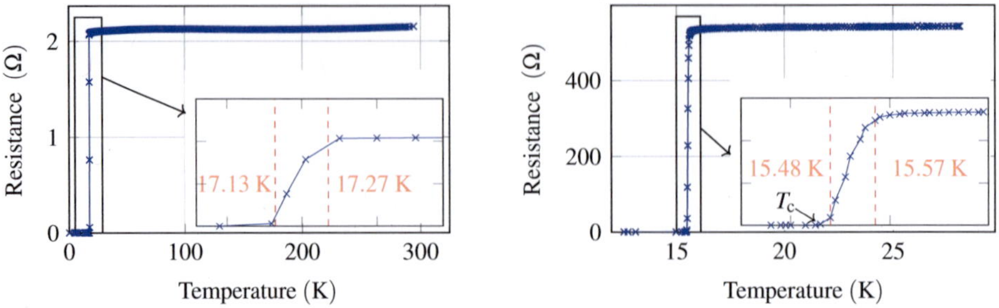

a) b)

Figure 4.11: Exemplary resistance measurement of NbN films on (a) MgO and (b) Sapphire substrates. For both measurements, the inset shows the same measurements magnified around the transition to the superconducting state. Furthermore, the $0.1 \cdot R_{20}$ and the $0.9 \cdot R_{20}$ temperatures are indicated in the insets by the red dashed lines.

This has been validated for thin films up to a thickness of 50 nm [117, 139]. However, this effect vanishes for films thicker than 50 nm. Since all films presented in this section exceed a thickness of 200 nm, the possibility of an additional AlN film is disregarded. The quality of a film is determined using the resistance measurement depending on temperature ($R(T)$) as depicted in figure 4.11 for two niobium nitride films, one deposited on MgO (4.11a) and the other one was deposited on a sapphire substrate (4.11b). The so-called residual resistance ratio *RRR* is the ratio between the room-temperature resistance and the resistance at 20 K

$$RRR = \frac{R_{300}}{R_{20}}. \tag{4.3}$$

This quantity can take values from < 1 which corresponds to films in the nitrogen phase to > 1 for the metallic phase. The critical temperature is read at the point, where the resistance has decreased into the noise level of the measurement system. The transition width is the temperature difference $T_{R=0.9 \cdot R_{20}} - T_{R=0.1 \cdot R_{20}}$ as shown in the insets of figure 4.11 and evaluates the homogeneity of the film. The resistivity ρ is derived from the resistance at 20 K and the geometry of the measured structure

$$\rho = R\frac{A}{l} \tag{4.4}$$

with $A = w \cdot t$ being the cross-section of the measured structure with a length l and film thickness t.

The most important parameter in this work is the energy gap of the superconductors. The optimization of the energy gap can be done by the optimization of the critical temperature

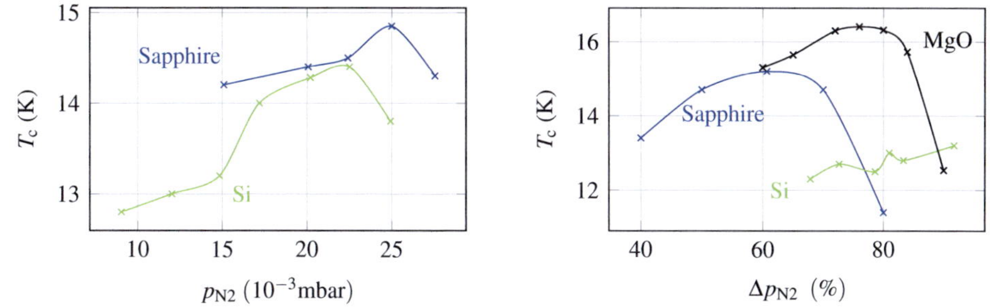

a) Si: $p_{Ar} = 2.4 \cdot 10^{-3}$mbar, $\Delta N_2 = 87$ %.
Sapphire: $p_{Ar} = 4.5 \cdot 10^{-3}$mbar, $\Delta N_2 = 67$ %.

b) Si: $p_{Ar} = 2.4 \cdot 10^{-3}$mbar, $p_{N2} = 0.42 \cdot 10^{-3}$mbar.
Sapphire: $p_{Ar} = 4.5 \cdot 10^{-3}$mbar, $p_{N2} = 1.5 \cdot 10^{-3}$mbar.
MgO: $p_{Ar} = 1.6 \cdot 10^{-3}$mbar, $p_{N2} = 0.59 \cdot 10^{-3}$mbar.

Figure 4.12: Optimization of the critical temperature of NbN films on different substrates. (a) The critical temperature of NbN films deposited on silicon and sapphire substrates, depending on the injected nitrogen content into the sputtering atmosphere. (b) The nitrogen consumption dependent variation of the critical temperature for NbN films on silicon, sapphire and MgO substrates. The respective constant parameters for each substrate and deposition are given in the captions of the subfigures.

(see equation (3.8)). Figure 4.12a depicts the measured values of the critical temperature of NbN films on different substrates depending on the nitrogen content in the total gas mixture. The argon pressure and the nitrogen consumption were kept constant, which was realized by adjusting the deposition current to reach the particular consumption for each nitrogen content. The NbN films on Sapphire were deposited at a heater temperature of 775°C and reached a maximum $T_c \approx 14.7$ K at a nitrogen content of 25 %. Furthermore, silicon-based substrates were investigated, in this case, the deposition was done without intentional heating by the heater. The two substrate types were oxidized silicon and single-crystal silicon [111]. The investigation thereof revealed only negligible differences between the NbN parameters deposited on these two substrates. Therefore, figure 4.12a only shows the results obtained on Si[111] substrates. The maximum critical temperature was achieved at a nitrogen content of 23 %.

Table 4.1: Overview of the achieved superconducting and electric properties of the NbN films on various substrates.

Substrate material	T_c(K)	ΔT_c (K)	ρ_{20} (μΩcm)	Deposition temperature (°C)
MgO	17	0.14	40	775
Sapphire	15.7	0.09	180	775
SiO2	14.5	0.1	145	RT
Si111	14.4	0.1	145	RT

The results of the subsequent optimization step is presented in figure 4.12b. In this optimization step, the nitrogen consumption is varied, keeping the argon and nitrogen content constant, where the optimum argon to nitrogen ratio from the previous optimization step for each substrate type is used. The nitrogen consumption is varied by a variation of the deposition current. Obviously, each curve exhibits an optimum nitrogen consumption except for the NbN films on the silicon based substrates. These films have been deposited up to a nitrogen consumption of $\approx 92\ \%$ after which the critical temperature drops dramatically. The NbN films, deposited on MgO substrates, exhibit the highest critical temperature followed by the films on sapphire substrates.

By this procedure, the critical temperature of the NbN films was optimized on every substrate separately, obtaining the maximum critical temperature of the NbN films for each substrate. The achieved NbN film-parameters are summarized in table 4.1.

Surface roughness

The second major optimization criterion of the NbN films is the surface roughness. Since the NbN film growth on MgO was expected to be monocrystalline, the surface roughness was not measured. Pure silicon was also disregarded, since it proved to result in slightly lower critical temperatures of the NbN films than oxidized silicon substrates. The films on oxidized silicon are deposited at room temperature, since the SiO_2 has an amorphous crystalline structure, which is not capable of enabling a monocrystalline growth of a NbN film. The surface roughness is expected to be strongly influenced by the deposition rate. In particular, higher deposition rates should result in larger surface roughness [130]. Figure 4.13 depicts the result of this investigation concerning the critical temperature of the resulting NbN films along with SEM images. The critical temperature exhibits a slight variation with the total pressure, however these changes are less than 500 mK. SEM images were taken in order to obtain a first indication of the surface structure which reveal a significant influence of the deposition current. High deposition currents result in SEM images with sharp, distinct features with a large contrast. The intensity of these features decreases strongly with decreasing deposition currents. An example of three SEM images is given in figure 4.13. The deposition parameters for these three films is summarized in table 4.2.

Atomic force microscope (AFM) measurements were conducted in order to quantify and validate the observations made by the SEM images. In fact the root mean square (RMS) surface roughness of the films increase significantly form 1.1 nm at $I_{\text{dep}} = 391$ mA to 1.5 nm at $I_{\text{dep}} = 560$ mA. This increase of almost 50 % in roughness for an increase in deposition

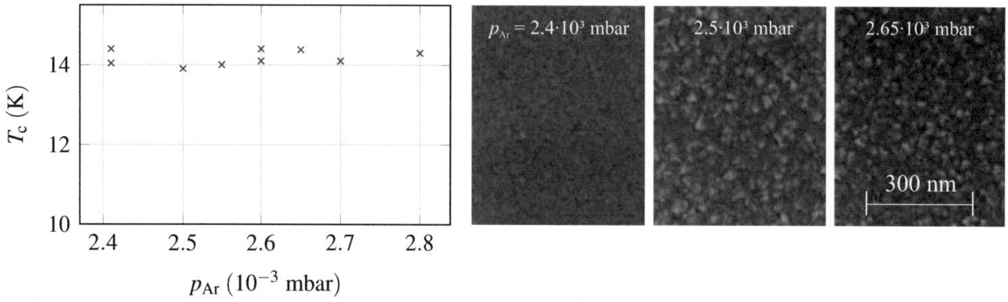

a) b)

Figure 4.13: Pressure scaling of NbN deposition on SiO_2 substrates. The argon pressure was varied from $2.4 \cdot 10^{-3}$ to $2.65 \cdot 10^{-3}$mbar. The SEM images in b) depict the surface of the resulting NbN films for different argon pressures. Data adopted from [Boh14].

rate of 50 % verified the assumption that a smaller deposition rate yields a lower surface roughness.

The same procedure can be applied to the deposition of NbN films on sapphire substrates. In this case, the variation in deposition current and argon pressure was selected in a wider range. Therefore, a scaling parameter P was introduced. This parameter is used to scale the argon and nitrogen pressure,

$$p'_{\mathrm{Ar}} = P \cdot p_{\mathrm{Ar},0}, \quad \text{with} \quad p_{\mathrm{Ar},0} = 1.4 \cdot 10^{-3}\ \mathrm{mbar} \tag{4.5}$$

$$p'_{\mathrm{N}_2} = P \cdot p_{\mathrm{N}_2,0}, \quad \text{with} \quad p_{\mathrm{N}_2,0} = 0.47 \cdot 10^{-3}\ \mathrm{mbar} \tag{4.6}$$

keeping a constant ratio of argon to nitrogen partial pressures. The nitrogen consumption is kept constant by adjustment of the discharge current. Smaller deposition currents include a decreased deposition rate. The result is depicted in figure 4.14a. The scaling parameter was varied in the range of $1 \leq P \leq 3$ and for each of these values, a variation in nitrogen consumption was conducted. Clearly visible is the accumulation point of the maximum critical temperature values around 65 %.

Table 4.2: Deposition parameters for the NbN films on SiO_2 substrates shown in figure 4.13

Film	p_{Ar} (mbar)	p_{N2} (mbar)	Δp_{N2} (%)	I_{dis} (mA)	V_{dis} (V)	r_{dep} (nm/min)
1	$2.4 \cdot 10^{-3}$	$0.61 \cdot 10^{-3}$	86	391	273	7.1
2	$2.5 \cdot 10^{-3}$	$0.73 \cdot 10^{-3}$	86	560	256	10
3	$2.65 \cdot 10^{-3}$	$0.77 \cdot 10^{-3}$	86	620	255	11.3

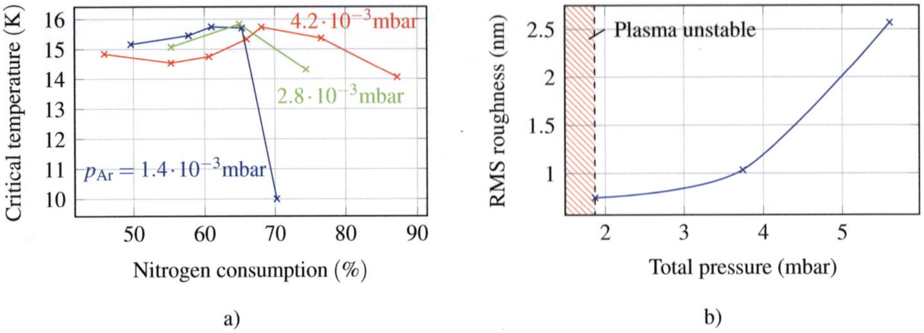

Figure 4.14: Total-pressure scaling of the NbN deposition process on sapphire substrates. (a) The critical temperature depending on the nitrogen consumption for different argon pressures. Along with the argon pressure, the nitrogen pressure was scaled with the same factor, keeping a constant p_{N2}/p_{total} ratio of 33 %. (b) The results from AFM measurements on the different NbN films deposited with the optimum nitrogen consumption but different total pressures.

Increasing the nitrogen consumption beyond this point results in a decrease of T_c. The higher the total pressure, the shallower the decrease in T_c after exceeding the point of maximum critical temperature. Within 3 %, the maximum critical temperature is determined by the nitrogen consumption, independent on the scaling parameter. The slight deviation might stem from the fact that for the low deposition pressures, the accuracy of the pressure sensors mounted in the system is already in that order of magnitude. Figure 4.15 depicts SEM images which, similar to the behavior on SiO_2, exhibit sharp crystalline structures on films realized at high pressures, which vanish for low process pressures. This is confirmed by AFM measurements of these films, which revealed the expected behavior of decreasing surface roughness of the 200 nm thick films with decreasing deposition rate. This is depicted in figure 4.14b. The lowest RMS roughness was achieved at a total pressure of $p_{total} = 1.87 \cdot 10^{-3}$ mbar, which is the lowest pressure, where the plasma in the deposition chamber is still stable. The deposition rate in this case is only $r_{dep} \approx 0.05$ nm/sec, which makes the deposition of an entire trilyer with 200 nm thick electrodes very time consuming. The lower limit for the surface roughness is thus given by the plasma stability. The optimized deposition parameters are summarized in table 4.3. The span given for the deposition current and voltage is due to a necessary adjustment of the deposition current over a long time span. This adjustment is most likely due to target erosion and was set in order to maintain the targeted nitrogen consumption.

The optimization of NbN films is discussed, revealing superior critical temperatures on MgO substrates as expected. The NbN film quality was also optimized on sapphire and SiO_2 substrates, yielding acceptable results on the silicon based substrates. The maximum critical temperature of the NbN films deposited on sapphire yielded only ≈ 1 K lower T_c as

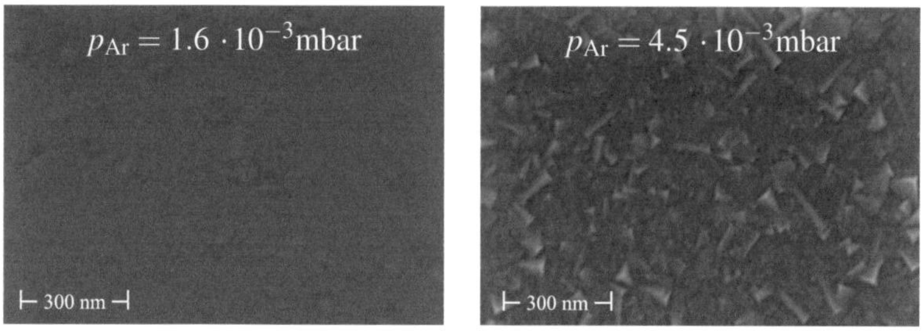

a) b)

Figure 4.15: SEM images of the surfaces of NbN films deposited with different scaling parameters P. The film deposited with $p_{Ar} = 4.5 \cdot 10^{-3}$mbar clearly exhibits sharp crystallite-like structures on the surface. These crystals almost completely vanish when approaching $p_{Ar} = 1.6 \cdot 10^{-3}$mbar.

compared to the films obtained on MgO substrates. The surface roughness was optimized as well, leading to small deposition rates which resulted in a surface roughness $r_{RMS} < 1$ nm on sapphire substrates. These values seem appropriate to fabricate Josephson junctions with these films.

For the fabrication of SIS trilayers, the insulating AlN barrier is deposited on top of the ground electrode. This will be explained using the example of sapphire substrates with elevated heater temperatures. However, for the room temperature deposition, the procedure is analogue except for the cooling down of the sample prior to unloading. Before opening the gate valve between the NbN and the AlN chamber, the NbN chamber is evacuated again. Then, the gate valve is opened and the entire heater with the sample on top is moved to the AlN chamber. Upon reaching the position under the aluminum target, the process gases for the AlN deposition are introduced and the deposition is started by igniting the plasma and opening the shutter. The process parameters for the AlN deposition have been optimized earlier [117] and have not been subject to optimization during this work. The set of process parameters used the AlN thin film deposition is given in table 4.3. The deposition of the counter electrode is done after returning the heater with the sample back to the NbN chamber, where the identical deposition parameters are used as for the ground electrode.

After cooling down of the heater, the sample holder can be transferred to the load lock for unloading. Patterning these samples with the fabrication process described above, enables the measurement of the single junctions. Some of the characterization junctions presented in this chapter were realized using a different patterning process, which allowed a shorter turnaround time. This process is described in [117, MBV$^+$17], however, is irrelevant for the determination of the junction quality. The junction properties, achieved by the optimization presented here, are presented in the next section.

Table 4.3: Overview of the deposition parameters for the deposition of NbN films on Sapphire substrates.

Material	p_{Ar} (mbar)	p_{N2} (mbar)	I_{dep} (mA)	V_{dep} (V)
NbN	$1.4\cdot10^{-3}$	$0.47\cdot10^{-3}$	70-100	530-540
AlN	$2\cdot10^{-3}$	$1\cdot10^{-3}$	200	240

4.3.2 Junction properties after the optimization of NbN films

The deposition parameters of the optimized films shown in the previous section are used to deposit different trilayers. The obtained Josephson junctions are presented and discussed in this section. As a benchmark, the realized junctions on MgO substrates are presented as well.

Due to the high critical temperature of the NbN films on MgO substrates, the junctions realized from these layers are expected to exhibit high energy gaps. From the current-voltage characteristic shown in figure 4.16, the gap voltage can be extracted as $V_{gap} = 6$ mV which is a record value for NbN-based Josephson junctions with AlN barriers. The ratio of sub-gap to normal state resistance also shows high quality since $R_{sg}/R_N = 26 > 10$, however, at a low critical current density of only $j_c = 0.14\,\mathrm{kA/cm^2}$. Nonetheless, the quality of this junction does serve as a benchmark due to the high gap voltage, the small gap smear and the low leakage. The critical current modulation with the magnetic field is shown in figure 4.16b, and follows the Fraunhofer pattern (red line), which demonstrates the homogeneous current density distribution across the junction.

The room-temperature deposited trilayers on SiO_2 substrates are expected to result in lower gap voltages due to the lower critical temperatures. According to the expectation, the highest

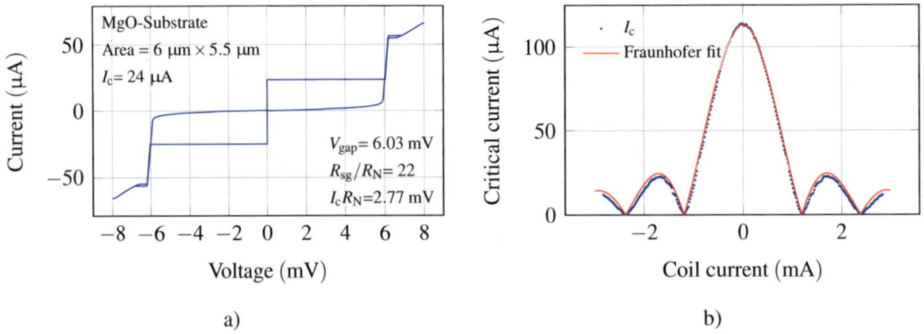

Figure 4.16: (a) Current-voltage characteristic of a Josephson junction on a MgO substrate. (b) The measurement of the critical current modulation by a magnetic field. The measured values closely follow the Fraunhofer fit, indicating a homogeneous critical current density distribution. Data adopted from [Non16].

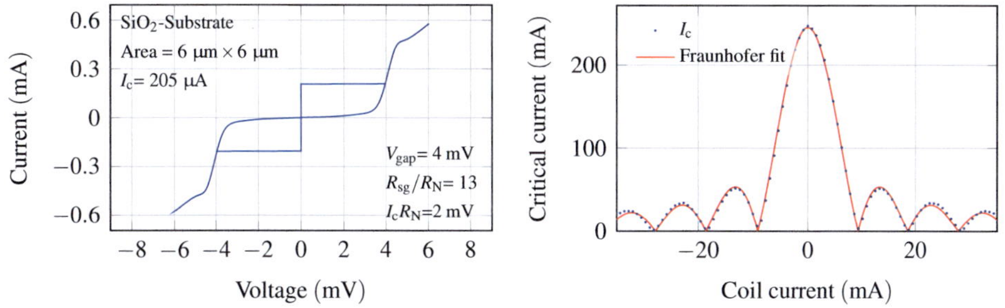

a) b)

Figure 4.17: Current-voltage characteristic of a Josephson junction deposited on a SiO_2 substrate. (b) An exemplary measurement of the magnetic modulation of the critical current. The measurement almost perfectly follows the Fraunhofer fit.

measured gap voltage of the trilayers on these substrates is $V_{gap} = 4\,\text{mV}$. An example of a current-voltage characteristic is depicted in figure 4.17a, showing that also the gap smear is significantly larger compared to the MgO benchmark. The junction however exhibits a low leakage at a slightly higher critical current density of $j_c = 0.6\,\text{kA/cm}^2$, which is still a rather low value. Concerning the current density distribution, this junction shows excellent properties, since the measured $I_c(H)$ modulation almost perfectly follows the Fraunhofer fit, merely deviating at higher magnetic fields.

In order to increase the low critical current density, the deposition time of the AlN barrier was decreased, resulting in thinner barriers. The obtained dependence of the R_{sg}/R_N-ratio in relation to the critical current density is depicted in figure 4.18. As expected, larger critical current densities exhibit larger leakages due to the thinner barriers. For low j_c values, the

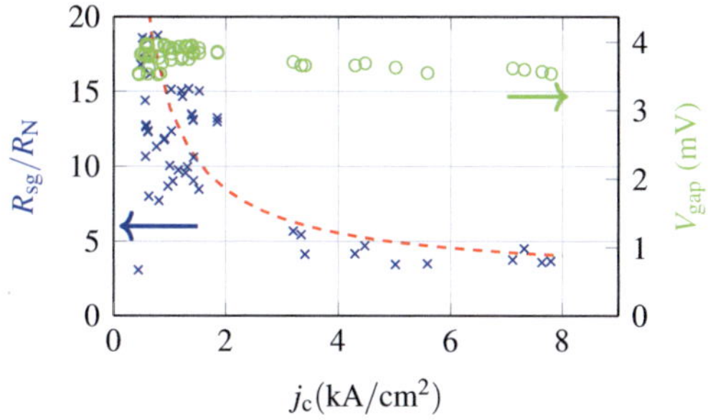

Figure 4.18: Overview of the realized quality parameters of junctions on SiO_2 depending on the critical current density. Higher j_c-values lead significantly decreased R_{sg}/R_N-ratios which is illustrated by the red dashed line. The graph also contains the measured gap voltage, which only shows a minor decrease for higher j_c-values.

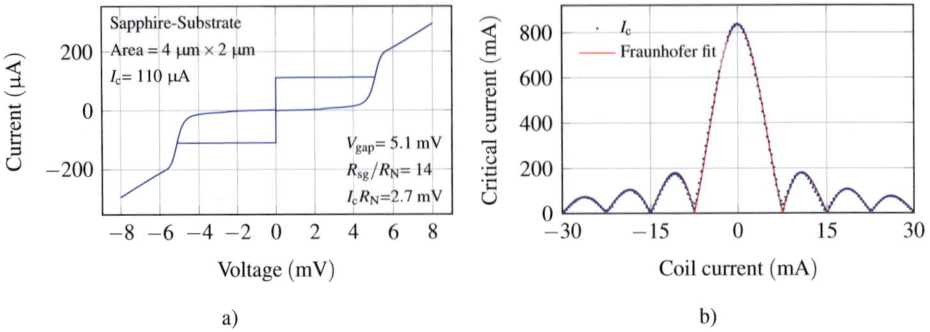

Figure 4.19: (a) Current-voltage characteristic of a junction on a sapphire substrate. (b) The modulation of the critical current by an applied magnetic field perfectly follow the Fraunhofer fit, indicating an excellently homogeneous critical current distribution across the junction.

leakages exhibit acceptable values, but rapidly decrease to $R_{sg}/R_N \leq 4$ for higher critical current densities. The green circles in figure 4.18 represent the measured gap voltages of the junctions, which should not vary with the critical current density. According to the data, the junctions with higher critical current density exhibit a slightly lower gap voltage. This may however be due to a long term deposition parameter drift. For a definitive correlation, more data are required.

The trilayers deposited on sapphire expectantly exhibit a higher quality than the ones realized on SiO_2. An example of a current-voltage characteristic is also shown for sapphire in figure 4.19a. The gap voltage obtained from this junction is as high as $V_{gap} = 5.1$ mV, which is already comparable with literature values of NbN-based junctions with AlN barriers deposited on MgO substrates. The R_{sg}/R_N value demonstrates a high quality barrier at a critical current density of already $j_c = 1.4\,\mathrm{kA/cm^2}$. The $I_c(H)$ modulation shows an excellent Fraunhofer pattern, proving an excellent homogeneity of the barrier. This result was only possible with a certain barrier deposition procedure.

Influence of the barrier deposition Figure 4.20 shows current-voltage characteristics of two Josephson junctions which were deposited with identical electrode deposition parameters, yet, the barrier deposition conditions are different. As described above, for the barrier deposition, the heater is moved to the AlN chamber where a low base pressure is present in both chambers. Introducing the process gases is done by opening the needle valves and closing the throttle valve of the turbomolecular pump. The deposition was usually not started before a certain settling time $t_{settle} = 5$ min was elapsed in order to guarantee a stable and constant atmosphere for the deposition. However due to this long settling time the surface of

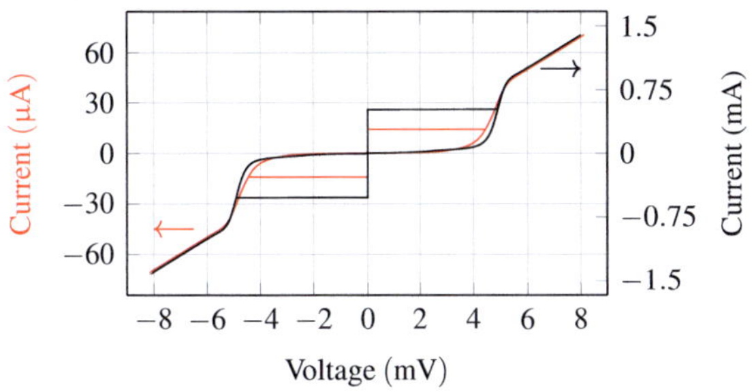

Figure 4.20: Comparison of the current-voltage characteristics of two Josephson junctions with identical deposition parameters for the electrodes. The gas settling times and the AlN deposition times were varied for these junctions from $t_{settle} = 5$ min and $t_{dep,AlN} = 25$ sec shown by the red curve to $t_{settle} = 15$ sec and $t_{dep,AlN} = 35$ sec, which is shown by the black curve.

the lower electrode is expected to deteriorate due to the high temperatures and the presence of nitrogen [127]. An example is given by the red curve in figure 4.20. The AlN deposition time in this case was 25 sec. A significant improvement has been achieved by reducing the settling time dramatically down to $t_{settle} = 15$ sec. A junction realized with such short settling time and a deposition duration of 35 sec is represented by the black curve. The comparison of the quality parameters is given in table 4.4.

The barrier was deposited with identical pressures and deposition currents. The difference between the qualities is significant. Apparently, the deposition time of AlN is not dominant for the critical current density determination, since the shorter deposition time does not manifest itself in a larger critical current density. Furthermore, the gap voltage seems to suffer from the long settling time and the tunneling is less ideal value for longer settling times. The fact that the leakage is lower for the increased settling time is not surprising, since j_c is significantly lower. Observing from the current-voltage characteristic, the gap smear is reduced for shorter settling times. Obviously, a long gas settling time results in a massive deterioration of the junction quality parameters. Therefore, the gas settling time has been kept at the low values of 15 sec for the following trilayers.

Table 4.4: Comparison of the quality parameters of NbN/AlN/NbN junctions on sapphire substrates with different gas settling times

t_{settle}	$t_{dep,AlN}$ (sec)	I_{dep} (mA)	j_c (A/cm^2)	V_{gap} (mV)	R_{sg}/R_N	I_cR_N
5 min	25	200	39	4.6	18	1.5
15 sec	35	200	1435	4.9	13	2.7

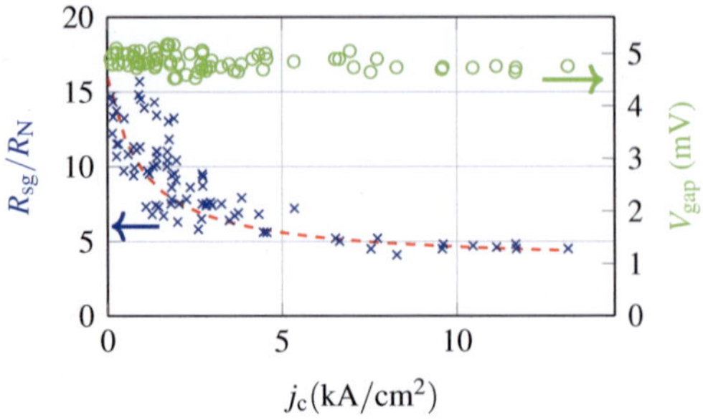

Figure 4.21: Overview of the V_{gap} and R_{sg}/R_N-ratio of junctions realized on sapphire substrates depending on the critical-current density. The R_{sg}/R_N-ratio exhibits a strong decrease for higher j_c-values, which is indicated by the red dashed line.

A variety of trilayers was realized with a deposition time ranging from 25 to 32 sec in order to increase the critical current density. So far, no unambiguous dependence of the critical current density to the barrier deposition time could be derived. This is most likely due to timing and pressure uncertainties of the deposition process. Figure 4.21 depicts the measured results of these trilayers. Again, for low j_c values, the junctions exhibit $R_{sg}/R_N > 10$, which decreases down to ≈ 5 for higher j_c values. Nevertheless, it was possible to demonstrate junctions with critical current densities as high as 13 kA/cm^2. The gap voltage should not be affected by varying j_c. In fact, the value appears to be independent of the critical current density however, is subject to variation of ~ 15 %. This variation occurs although all trilayer electrodes were deposited with the identical deposition parameters. Apparently, fluctuations of the deposition parameters must be present, which, in the current setup of the deposition system, can neither be observed nor avoided. This emphasizes the sensitivity of the NbN films to the deposition conditions. In order to reduce or even remove these fluctuations, the deposition system needs to be further improved.

Conclusion

The development of high-quality NbN-based Josephson junctions is presented in this section. This includes the optimization of the critical temperature, as well as the minimization of the surface roughness of the single NbN thin films on different substrates. Due to the complexity of the reactive sputtering process, distinct optimization criteria have to be established, that enable the optimization of a certain parameter without the deterioration of another parameter. It has been found, that the optimization of the critical temperature can be achieved by optimizing the nitrogen injection in conjunction with the nitrogen consumption. The surface roughness can then be optimized by keeping these parameters constant and reduce the total pressure, by scaling each pressure by the same factor. This leads to the observation of a reduced surface roughness with a decreased total pressure and hence, a decreased deposition rate. This optimization was conducted until the limits of the deposition system. The deposition parameters for the AlN barrier were adopted from [117], however, the procedure of the barrier deposition had to be modified due to interdiffusion issues at the elevated deposition temperatures, when using sapphire substrates. Therefore, the gas setup times were reduced. The optimization of the deposition process resulted in high quality Josephson junctions that fulfill the required needs for the realization of an integrated receiver, operating at frequencies higher than 700 GHz. As opposed to the pure Nb-based technology, the electrode thickness has increased by more than twice, which is why the patterning process has to be optimized and the limits need to be redetermined. This will be discussed in the following section.

4.4 Fabrication and patterning limits

With the Josephson junction quality parameters obtained from the measurements shown in the previous section, the basis for high-frequency devices is achieved. However, in order to establish design rules for the RF design, the patterning limits have to be quantified. These limits are different from the ones obtained with niobium-based trilayers, since the electrode thickness has more than doubled for NbN electrodes.

Again, the junction miniaturization plays an important role in order to keep the capacitance of the mixer junction as small as possible. Therefore, junctions with different areas have been realized and measured. The result of the measurements is depicted in figure 4.22. The junctions were designed down to areas as small as $0.5 \times 0.5\ \mu m^2$, which could all be measured successfully. The fabrication process is hence capable of realizing junctions with such small areas, which at a critical current density of $6\,kA/cm^2$ already exhibit a normal state resistance of $150\,\Omega$. This makes a further miniaturization impractical, since this high normal state resistance is already challenging for the impedance matching of the mixer junction.

Furthermore, the wiring layer needs to be investigated, since the microstrip lines are realized in this layer. Since the mixer junctions can exhibit high normal state impedances larger than $150\,\Omega$, which need to be matched to the low output impedances of the FFO, the transmission line network needs to be capable of achieving the required transformation impedances. In order to achieve high impedances, the width of the microstrip line needs to decrease or the dielectric thickness needs to be increased. In figure 2.4b, the achievable impedance ranges for two dielectric thicknesses are shown. The case for large dielectric thickness ($t_{diel} = 600$ nm) and the planar case ($t_{diel} = 200$ nm). The minimum wiring layer thickness is given by its London penetration depth ($\lambda_L \approx 200$ nm).

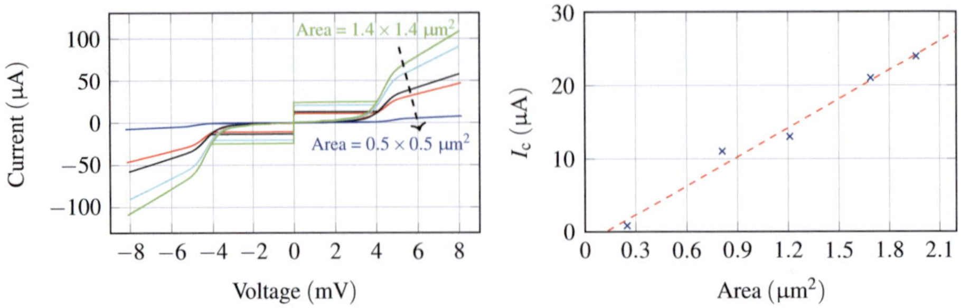

Figure 4.22: (a) Current-voltage characteristics of Josephson junctions fabricated on the same chip with different sizes ranging from $1.4 \times 1.4\ \mu m^2$ down to $0.5 \times 0.5\ \mu m^2$.The quality parameters are $V_{gap} = 4.6\,mV$, $j_c = 6\,kA/cm^2$, $I_c R_N = 0.8..1.7\,mV$ and $R_{sg}/R_N = 16..7$. (b) The critical current vs. JJ area along with a linear fit indicating a good reproducibility of the fabrication process.

The planar case makes it difficult to reach high impedances, since deep sub-micrometer structures are required. A 600 nm thick dielectric would lead to at least 200 nm steps which required a 400 nm thick wiring layer. In both of these cases, there is no margin for possible thickness fluctuations, which is why, the dielectric thickness of the wiring layer for the NbN-based technology was investigated assuming a 400 nm thick dielectric with a 400 nm thick wiring layer. Two possible patterning techniques can be applied, since the wiring layer is the last layer in the fabrication sequence. It can be patterned using a lift-off technique or it can be etched. Both techniques have been investigated, the result is depicted in figure 4.23. These SEM images show the cross-section of possible microstrip lines. Note the different scalings. The substrate is seen in the bottom part of the pictures, where the ground electrode is deposited. On top of the ground electrode, the 400 nm thick dielectric is deposited. The microstrip is on top of the dielectric. As can be seen, the patterning using lift-off results in very shallow edges, which are not beneficial for transmission lines. Furthermore, these edges

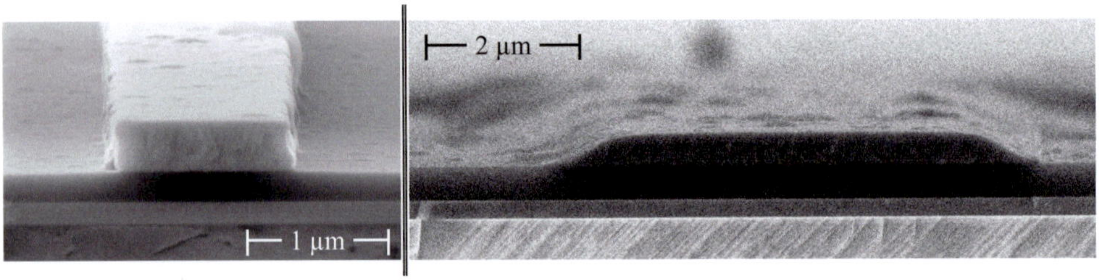

Figure 4.23: Tilted SEM images of wiring layer stripes. The left image depicts a NbN stripe in the wiring layer structured by etching to a width of 1 μm. The right image shows the SEM image of a NbN stripe patterned by lift-off exhibiting very shallow edges.

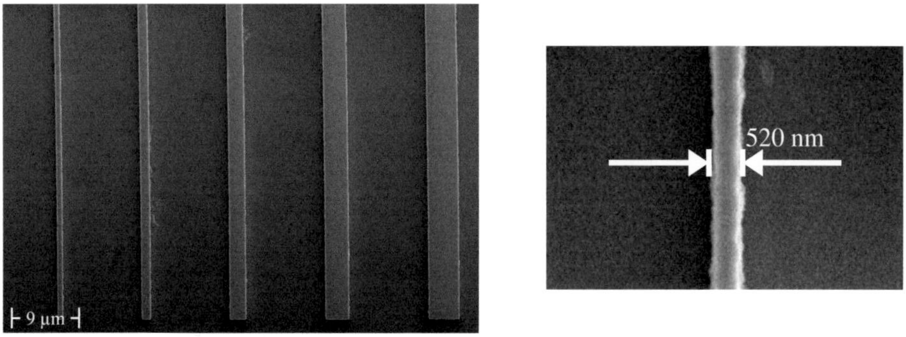

a) b)

Figure 4.24: (a) An SEM image of a set of stripes in the wiring layer with successively reduced width from the right to the left. The patterning technology allows a linewidth as narrow as 520 nm as indicated in (b).

also limit the miniaturization of the width of the transmission line. Therefore, the etching is preferred, as it shows steep edges.

The wiring layer patterning by etching was further investigated concerning the minimum achievable linewidth. A variety of 400 nm thick lines has been realized with decreasing width. The result is depicted in figure 4.24. The smallest width with acceptable edge roughness turned out to be 520 nm. This is thus the lowest linewidth that can be realized, which already exhibits a significant amount of surface roughness. This needs to be considered when such narrow linewidths are realized, since the surface roughness might introduce a significant amount of RF losses.

Conclusion

The patterning limits with respect to the needs of an integrated receiver are investigated in this section. These requirements mostly stem from the RF properties of the various devices. First, the miniaturization of the junction is of utmost importance. The fabrication process yielded square shaped junctions with an area of only $500 \times 500\ \text{nm}^2$. For the critical current densities achieved so far, such small junctions already exhibit a high normal state resistance, which is impracticable for impedance matching. These junction areas are thus small enough. Another major important optimization criterion is the width and shape of the wiring layer, which will be realizing the microstrip transmission lines. Therefore, the patterning techniques were compared. The etched structures clearly yielded better results which manifests itself by steep edges in contrast to the shallow edges obtained by lift-off. Therefore, the lithography and etching of the wiring layer were investigated with respect to the minimum achievable linewidth, resulting in lines as narrow as 500 nm. The quantified patterning limits are used as design rules for the simulations, which will be discussed in the following chapter.

5 Simulation of integrated receivers

Due to the optimization described in the previous chapter, the DC-properties of the materials and the structural limits are determined. However, in order to develop the structures that exhibit the necessary RF properties, simulations have to be conducted, investigating and optimizing the various structures needed for the operation of an integrated receiver. For the simulations in this work the software "CST Studio Suite" [140] was employed. This software package consists of multiple different sub-tools, and contains the "CST Microwave Studio" (MWS), which is employed for a field simulation of 3D objects. Furthermore a schematic tool is available in the package, which operates similar to a spice program. Various elements can be inserted and connected to each other such that the resulting circuit can be calculated. These elements can be either discrete parts such as resistors or capacitors, or blocks that comprise the results of 3D field simulations. Using these tools, the required structures of an integrated receiver can be optimized and assembled to the complete circuit.

5.1 General overview and challenges

MWS is a software tool that performs a full 3D RF-analysis of a given structure by solving the Maxwell equations. The structure under investigation is therefore discretized and the equations are solved in the resulting sampling points. Performing the simulation is done by assuming a certain excitation signal, which is applied to the structure and the response is calculated from which the $\underline{S}$-parameters can be derived. The result of the calculations is therefore the $\underline{S}$-parameter matrix. For complex structures that are large compared to the wavelength of the investigated signal, the calculations are likely to require simulation times that are of the order of days and the memory is intensely used. This is especially challenging for integrated receivers, since the complexity of the system is significantly higher as compared to mixers without integrated flux-flow oscillators. This can be seen in figure 5.1, which depicts the schematic representation of an integrated receiver split into different functional groups. The integration of the flux-flow oscillator leads to the necessity of additional impedance matching elements as well as a signal combiner on chip.

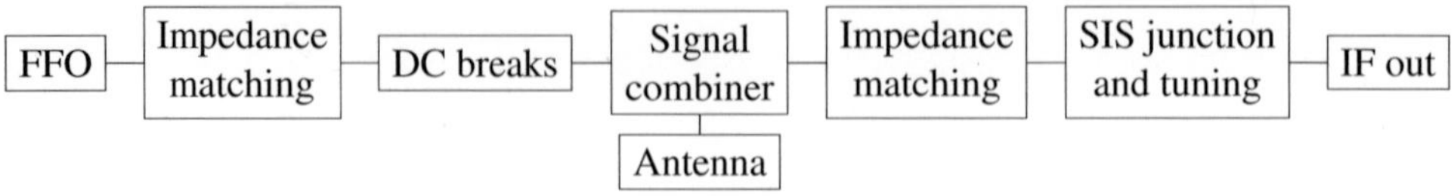

Figure 5.1: Block diagram of an integrated receiver, which is subdivided into these blocks.

Additionally, since the mixer is not the only active device on chip, it has to be separated for DC from the second active device, the FFO, which is realized by DC breaks. The high complexity is the reason, why it is not possible to optimize the entire structure of a superconducting integrated receiver in a single 3D field simulation. The simulation can however be split into suitable sub-simulations, where each simulation contains one of the functional groups. The single blocks can afterwards be assembled in the schematic tool and the transmission characteristics of the entire circuit can be obtained. Thus, the single parts can be optimized separately but the total simulation result is still obtained. Care has to be taken about the selection of the splitting location. Ideally, the splitting point is located in a structure without geometrical variations, e.g. a matched transmission line.

The excitation of the structure is done using "Ports". A variety of these ports is available in MWS, where the correct port has to be selected for the particular purposes. A common port is the "waveguide" port, which is typically used for the excitation of a microstrip transmission line. The port size is selected such that the electromagnetic field distribution does not exceed the limits of the port. Using this type of port, all passive components of the required structures can be simulated. The active devices, i.e. the mixer and the FFO, cannot be simply inserted into the simulation as discrete element, since CST does not provide these elements. Instead, they need to be represented by ports, which have to be selected correctly.

Another challenge arises, when it comes to the material selection. MWS provides a large library, containing a variety of different conductors and dielectrics. However, there is no superconducting material available. Therefore and since the material parameters of the optimized films may deviate from literature values, the need for an agile material parameter definition in MWS arises. By applying Mattis-Bardeen theory, the material parameters for the different NbN films have been derived, the result is shown in figure 5.2. The blue line represents the lower electrode, the dashed blue line depicts the wiring layer and the green line shows the result for a 600 nm thick aluminum film as a reference. Figure 5.2a depicts the surface resistance per square ($\Omega/\square$), the surface reactance is depicted in figure 5.2b and is also given in ($\Omega/\square$). The values obtained for the NbN films of the lower electrode coincide well with data shown in [141]. However, it can be seen from the data shown in this graph, that the wiring layer exhibits a significantly lower quality than the lower electrodes resulting

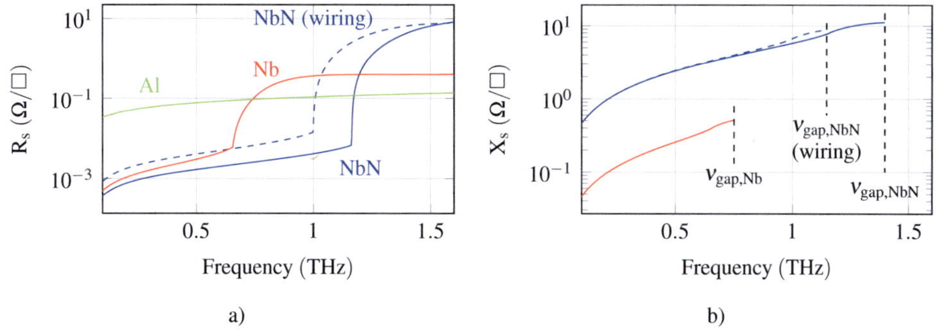

Figure 5.2: Surface impedances obtained by measuring T_c and ρ_{20} and evaluating the Mattis-Bardeen theory for 90 nm thick Nb and 200 nm thick NbN films. For NbN, two curves are shown: the ground electrode and the wiring layer. The green curve is the result of a measurement and calculation for a 600 nm thick Al sample for comparison. (a) The resulting surface resistance of the materials. (b) The respective reactances.

from the room-temperature deposition on an amorphous underground. This behavior of NbN films is known [28], which is why in future devices, the wiring layer material might be replaced with a different, low loss, high gap material such as NbTiN. The red curve depicts the surface impedance of pure niobium also extracted from $R(T)$-measurements with $T_c = 9.2$ K and $\rho_0 = 2.34$ µΩcm of a 90 nm thick film. Clearly visible is the lower gap frequency and the lower surface reactance of niobium. Although it is drawn as lines, the values shown in figure 5.2 are discrete points, since the solution of the integrals (compare equations (3.16) and (3.17)) are calculated numerically. These discrete values can be entered in MWS as "tabulated surface impedance". In order to obtain a continuous frequency dependent function of the impedance, MWS applies a fit to each R_s and X_s, mapping the impedance behavior as close as possible. The influence of especially the imaginary part of the NbN surface impedance is revealed noticeably, when regarding the transmission line characteristic impedance.

The dielectrics that need to be considered in the simulations are R-plane sapphire (Al_2O_3) and oxidized silicon as substrates and silicon monoxide as dielectric. The permittivity values of these dielectrics were adopted values from literature. In the simulations, these materials were considered lossless with the permittivities $\varepsilon_{r,Al2O3} = 11$, $\varepsilon_{r,silicon} = 11.9$ and $\varepsilon_{r,SiO} = 5$.

Based on these material parameters, the simulations can be conducted for the particular material systems. Due to the different London penetration depth, the layer thicknesses are different for niobium-based trilayers compared to trilayers with niobium nitride electrodes. The particular values are given in table 5.1. In section 4.4 the thickness of the dielectric was discussed concerning the planarization capabilities of the fabrication process.

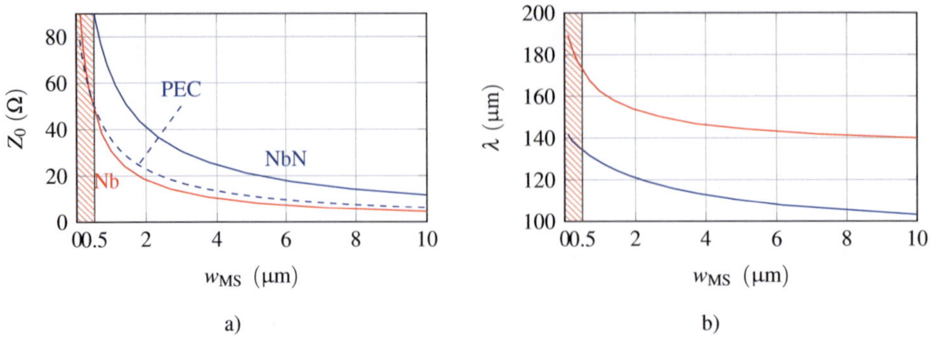

Figure 5.3: (a) Simulated characteristic impedances of microstrip transmission lines with the respective material parameters included in the simulation. The Nb-based transmission lines are simulated with a 200 nm thick dielectric, whereas the NbN-based transmission lines use a 400 nm thick dielectric. The dashed blue line represents the transmission line impedance of the NbN technology with PEC as conductor material. The simulations are all performed for a frequency of 700 GHz. The corresponding wavelength slope is depicted in (b). The hatched areas depict the patterning limits.

As apparent from table 5.1, the dielectric thickness does not coincide with the electrode thickness which was demanded for the planarization of the surface. It is essentially a trade-off between the requirements to the patterning of the wiring layer which would require deep-sub micrometer structures in the planar case and the steps that have to be covered by the wiring layer. The given values have been selected in order to achieve a range of characteristic impedance of the microstrip lines which is suitable for the RF circuit in an integrated receiver. All simulations presented in the following, have been done on the basis of the material parameters given in table 5.1.

Figure 5.3a depicts the simulated characteristic impedances of microstrip lines for the both, Nb- and NbN-based trilayers depending on the width of the microstrip. Clearly visible is the higher impedance range covered by the NbN-based microstrip lines. This is not only due to the lower capacitance resulting from the thicker dielectric, but also due to the larger kinetic inductance.

To demonstrate the need for the inclusion of the material parameters, figure 5.3a also contains the simulated characteristic impedance of a microstrip line, where the NbN has

Table 5.1: Overview of the layer thicknesses for devices fabricated from NbN/AlN/NbN and Nb/Al-AlOx/Nb trilayers.

Electrode material	bottom electrode	counter electrode	total SiO	wiring layer
Nb	90 nm	90 nm	200 nm	150 nm
NbN	200 nm	200 nm	400 nm	400 nm

been substituted by a perfect electric conductor. The striking difference in characteristic impedance of almost a factor of two clearly shows the importance of the material parameters in the simulation. The hatched red areas in the graphs indicate the patterning limit of the microstrip width of 520 nm. Consequently, the microstrip line with 400 nm dielectric can cover an impedance range up to $\approx 80\,\Omega$. According to equation (2.25), the wavelength also varies with the inductance and capacitance. In order to compare both technologies, figure 5.3b depicts wavelength over the microstrip width. The simulations were done in both cases for a frequency of 700 GHz. The wavelength is clearly larger in the case of the niobium technology. This has to be considered during design, since the wavelengths change with the characteristic impedances and the technology.

Conclusion

The concept of the simulation splitting is introduced in this section. This enables the optimization of the single components to a desired behavior. Therefore, the entire design is subdivided into single functional groups which are optimized separately. The schematic tool can be used to recompose the particular parts to obtain the response of the entire circuit. The surface impedance of the particular materials needs to be included in the simulations due to the strong influence on the wavelength and the characteristic impedance. Therefore, the material parameters are calculated, using the results of DC measurements of the various films and evaluating the equations obtained from the Mattis-Bardeen theory. This has to be done for each technology separately, since the material parameters as well as the film thicknesses differ significantly. Using these results, the various components can be optimized, which will be presented in the next section.

5.2 RF-development of the various parts

The development of the different functional blocks will be demonstrated in this section, where each part has a different set of requirements to fulfill. This includes and determines the port selection for each component.

5.2.1 FFO integration and impedance matching

The first part in the signal chain is the flux-flow oscillator. As mentioned above, this active component can not be implemented as a dedicated device, but needs to be modeled using a port. This becomes especially significant, since the FFO is tapered to a narrow width in order to achieve higher output impedances as shown in section 3.2.3.

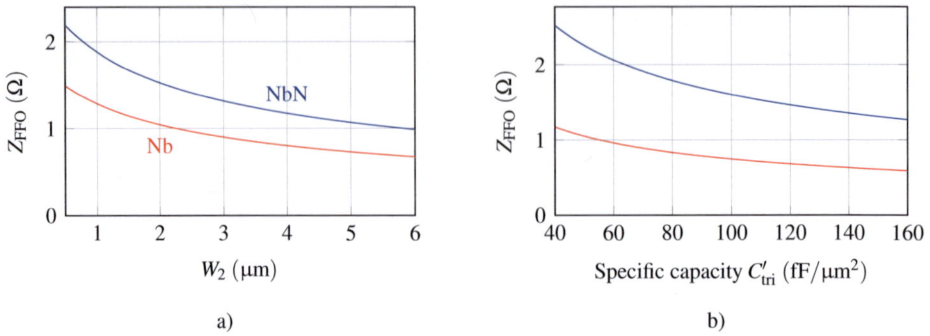

Figure 5.4: Calculation of the FFO output impedance for varying W_2 and capacitance C'_{tri}. (a) The dependence of the output impedance of the FFO for given specific capacity $C'_{tri} = 80\ \mathrm{fF}/\mu\mathrm{m}^2$ and $W_1 = 1\ \mu\mathrm{m}$ on the width of the idle region W_2 for both, the Nb as well as the NbN-based technology. (b) The variation of the FFO output impedance for a given geometry ($W_1 = 1\ \mu\mathrm{m}$, $W_2 = 1.5\ \mu\mathrm{m}$) depending on the specific capacity.

The output impedance of an FFO is typically in the range of $0.1\ \Omega < Z_{FFO} <$ few Ω. A microstrip line with such a small characteristic impedance would exhibit an extremely large, unpractical width. Therefore, the impedance matching component is directly attached to the FFO, transforming the low FFO output impedance to a more practical impedance. The practicability of this transmission line is mostly given by the subsequent DC breaks, which are not realizable using very wide transmission lines. On the other hand, very high impedances are not practical either, since the required narrow transmission lines increase the RF-losses and the structures are more sensitive to fabrication tolerances. In addition to that, the matching transformer between FFO and DC breaks would have to cover a large impedance span, which dramatically decreases the bandwidth. Therefore, an intermediate impedance section is introduced, which hosts the DC breaks and the T-junction. Due to the different material parameters, the impedance values of this intermediate impedance section were selected separately to 11 Ω corresponding to $w_{MS} = 4.5\ \mu\mathrm{m}$ for the Nb-based technology and 20 Ω, which corresponds to $w_{MS} = 5.68\ \mu\mathrm{m}$ in the NbN-based technology. Both of these widths proved suitable for the realization of the subsequent devices.

The FFO output impedance was derived in general in section 3.2.3. For the different fabrication technologies, this results in the impedances shown in figure 5.4. The influence of the width of the idle region W_2 on the FFO output impedance is shown in figure 5.4a. For these calculations, the trilayer capacitance is assumed with $C'_{tri} = 80\ \mathrm{fF}/\mu\mathrm{m}^2$ and a FFO output width $W_1 = 1\ \mu\mathrm{m}$ for both, the niobium-based as well as the NbN-based FFO. Since these two curves depict two different technologies, the values for the layer thicknesses and London penetration depths are used according to the particular technology. Clearly visible is the lower output impedance of the niobium-based FFO for a given geometry compared to the NbN technology. Figure 5.4b depicts the influence of the trilayer capacitance for a constant lateral geometry of the FFOs. Obviously, higher trilayer capacitances decrease the output impedance. Since the output impedance of the FFO is very low compared to the remaining network, lowering the output impedance is not beneficial for the impedance matching. Therefore, the impedance matching of the FFO does not benefit from higher j_c values.

As discussed in section 2.3.3, the impedance matching can be done using various elements, among which the most prominent are the $\lambda/4$ and the Chebychev transformer. Both flavors have been used in this work, a schematic top view on these structures is depicted in figure 5.5. The dark gray areas hereby depict the bottom electrode, the light gray areas represent the wiring layer. The excitation of the structure is done using a face port. This port is selected, since it exhibits a certain spatial extension, which can emulate the spatial extension of the flux-flow oscillator as opposed to a discrete port, which is only a point like excitation source. This enables the modeling of the discontinuous width at the transition of the FFO to the impedance transformer. In case of a port that covers the full width of the first transformation stage, no discontinuity is present, and the length of the first stage is $\lambda/4$. However, if the port width is selected according to the FFO output width, the resulting discontinuity in width results in an electrically longer length of the first transformation section. The effect can be seen in figure 5.5b. Due to the large width of the first transformation stage, this section appears electrically longer than its physical length, which is why it has to be shortened in order to achieve the desired electrical $\lambda/4$ length. Of course, the intensity of this effect is depending on the wavelength compared to the widths of the microstrip lines. The longer the wavelength, the lesser the shortening of the first stage section. In case of long wavelengths, the width of the FFO port becomes less important. However, for smaller wavelengths and a port that reflects the width of the FFO, the shortening becomes stronger. Therefore, the Chebychev transformer is limited in frequency. The shorter wavelengths at higher frequencies result in extremely short first sections of the transformer, which make it impracticable above certain frequencies. Since the first stage of a Chebychev multi-section transformer always exhibits a smaller characteristic impedance as a single stage transformer, a $\lambda/4$ transformer may however still be applied. The higher impedance corresponds to a narrower linewidth, hence the shortening effect is less prominent. For these schematic structures, figure 5.5 contains the scattering parameters of the different structures, simulated in different technologies where Nb is represented by the red lines and NbN is depicted by the blue lines. For these simulations, the FFO output impedance was assumed with $Z_{\mathrm{FFO,Nb}} = 0.4\ \Omega$ for the niobium technology and $Z_{\mathrm{FFO,NbN}} = 2\ \Omega$ for the NbN technology.

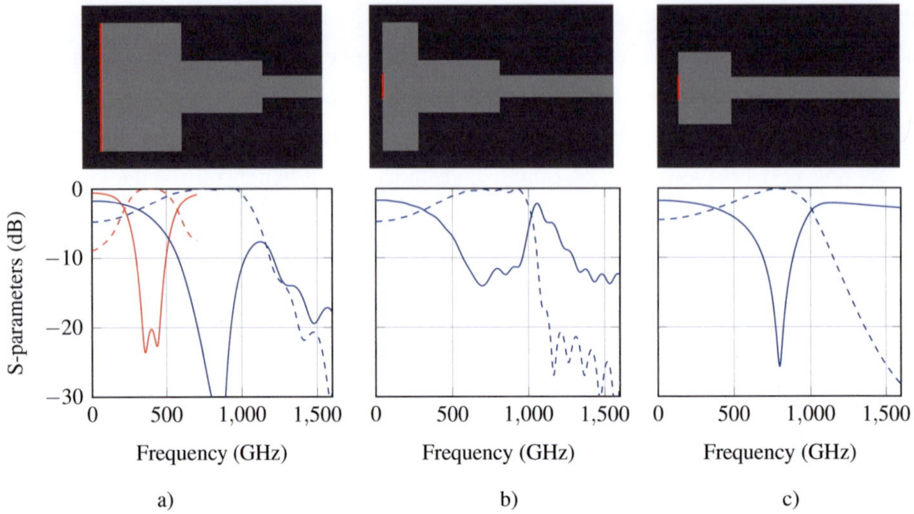

Figure 5.5: Various structures for the FFO impedance matching to the transmission line network. The structures are shown in the top part with the red lines indicating the excitation ports. The Chebychev matching transformer with a wide port (a) was simulated for 400 GHz using niobium and for 800 GHz with niobium nitride metalization. (b) and (c) depict a Chebychev and a quarter-wavelength matching transformer with a port width selected according to the FFO output width. The resulting S-parameters are depicted below the respective design images. The red line represents the case of Nb, whereas the blue lines show the simulation results with NbN. The graphs contain both, $|\underline{S}_{11}|$ as solid lines and $|\underline{S}_{21}|$ as dashed lines.

The output impedances of the transformers are the particular intermediate impedance values. For the design of the Chebychev matching transformer, the widths of the particular stages are selected according to the values given in the table [38], whereas the width of the single stage quarter-wavelength transformer can be calculated by (2.44) and the dependence given in figure 5.3a. The lengths of the different sections can be taken from the relation shown in figure 5.3b. The simulation result for a Chebychev matching transformer in Nb technology for a design frequency of 400 GHz is depicted in figure 5.5a as the red line. The solid line hereby depicts the $|\underline{S}_{11}|$ parameter, whereas the dotted line represents the transmission characteristics. Clearly visible is the ripple in the pass band of the $|\underline{S}_{11}|$ curve, but the matching with a reflection of less than -10 dB is achieved over a wide frequency range of 200 GHz. The figure also contains the simulation result of a Chebychev transformer in NbN technology optimized for 800 GHz excited with a wide port. The simulation of the Chebychev transformer using an excitation with the narrow port is shown in 5.5b, also optimized for 800 GHz. The matching is already distorted, the length of the first stage has to be shortened significantly. However, still a good matching can be achieved. The simulation of a $\lambda/4$ single stage trans-

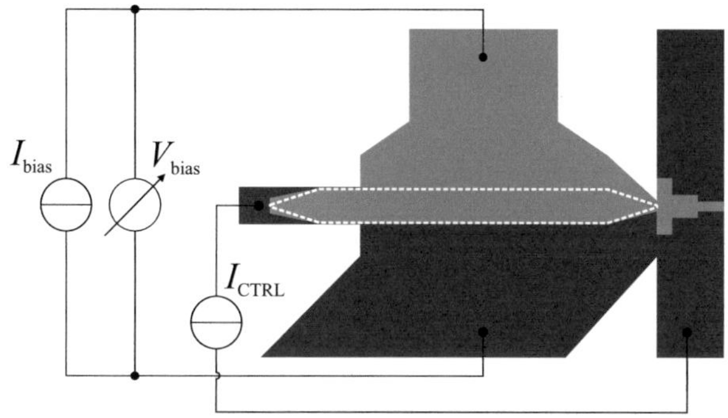

Figure 5.6: FFO design with connectors attached to the microstrip line network. The FFO output is at the right boundary of the flux-flow oscillator (depcited by the white dashed line). The matching transformer is attached directly to the FFO output and connected to the microstrip network towards the right.

former is depicted in figure 5.5c. Compared to the Chebychev transformer excited with a narrow port, the bandwidth is lower, but the matching at the design frequency is actually better. For a particular application, the appropriate matching transformer has to be selected.

Figure 5.6 depicts the full setup of the FFO connected to the impedance matching, in this case a Chebychev transformer. Again, the dark gray areas depict the bottom electrode, whereas the light gray areas represent the wiring layer. The junction is depicted as the dashed white line on the overlap of both NbN layers. The tapering is indicated for the FFO itself as well as the electrodes, which are tapered in order to reduce the width of the idle region. Furthermore, this figure also contains the connections for the control line, which is realized in the bottom electrode. The current is injected in the bottom electrode parallel to the barrier at one end of the flux-flow oscillator and extracted at the other end, where the FFO is attached to the matching network. This current creates the necessary magnetic field for the FFO operation. Additionally, the FFO also contains an unbiased tail. This has been shown to be beneficial for the slope of the oscillator current steps [114]. From the matching transformer attached to the FFO, the radiation can travel along the microstrip transmission line to the subsequent device on chip.

5.2.2 DC breaks

Following on the FFO and the matching transformer, the second element group is the DC breaks. The purpose of these elements is to enable the simultaneous biasing of the FFO and the mixer. Since FFO and mixer are connected in parallel, the DC-connection between them needs to be interrupted, while maintaining a good RF transmission. In theory, it is sufficient to only split the DC connection in the wiring layer.

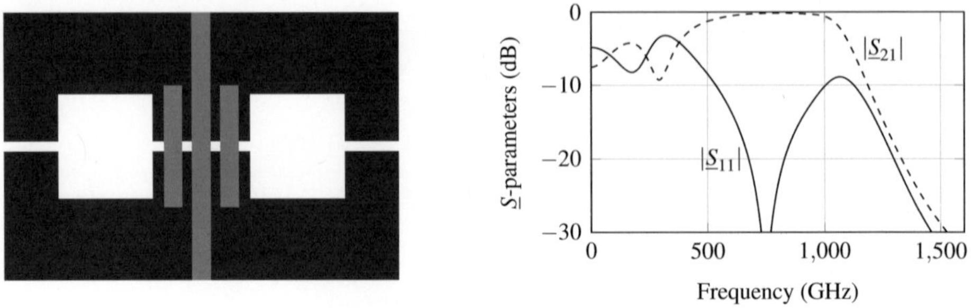

a) b)

Figure 5.7: (a) Schematic top view of a DC break in the ground electrode. The wiring layer (light gray) transverses the gap (white areas) in the ground electrode (dark gray), which blocks the DC-current. The RF signal on the other hand can pass the gap with negligible reflection losses. (b) An example of the simulation result of such a DC break targeted for 750 GHz.

However, this can lead to ground loops resulting in distortions, which is why a separation of both, the wiring layer as well as the ground electrode is beneficial. The idea for the design of the DC breaks is adopted from [142, 143], since it has proved superior performance over other approaches that have been simulated. For the excitation of such a device, waveguide ports are used. The optimization criteria are the maximized transmission in the passband along with a suppression of the DC-current transport.

The design of the ground electrode DC break is depicted in figure 5.7. A gap in the ground electrode ensures a high-Ohmic connection between FFO and mixer in this layer. The microstrip in the wiring layer crosses this gap with two additional conductor sections in parallel. These parallel conductors determine the center frequency of the signal transmission across the gap. The transmission level in the pass band is mostly determined by the width of the slit separating the bottom electrode. If the cut is realized as a simple slit of constant width, higher frequencies tend to exhibit slit modes that propagate along the slit and thus decrease the transmission to the intended output port. Therefore, the original design was altered, adding sections with wider slits, as can be seen in the design. The simulation results of the structure can be seen in figure 5.7b. Although the transmission is enhanced by the resonant parallel conductors, the $|\underline{S}|$-parameters reveal a large -10 dB bandwidth of more than 400 GHz and a good isolation at low frequencies. The roll-off in transmission is due to the increasing losses in the wiring layer material.

Figure 5.8a depicts the design of the DC break in the wiring layer. The wiring layer is split in order to achieve the required high-Ohmic connection for DC. However, to provide a high RF transmission, the wiring layer is provided with two short conductors in parallel to each other. The ground electrode exhibits a "U"-shaped cut as well in order to improve

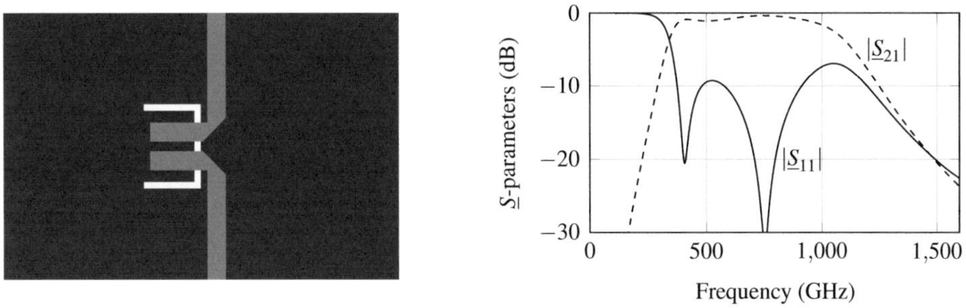

a) b)

Figure 5.8: Design of a DC break in the wiring layer. (a) The schematic top view with the dark gray representing the ground electrode and the light gray as the wiring layer. The white areas represent the gaps in the ground electrode. The RF coupling is granted by the parallel pieces of transmission line with the U-shaped cut in the ground electrode. The corresponding scattering parameters are depicted in (b).

the transmission. The center frequency of transmission is determined by the length of the conductors, whereas the cut in the bottom electrode significantly improves the transmission level. Figure 5.8b depicts the $|\underline{S}|$-parameters for this structure. The transmission of low frequencies is clearly suppressed. The designed structure exhibits a large −10 dB bandwidth along with a good matching in the pass band. The roll-off of the transmission is again due to the increasing losses in the wiring layer material.

5.2.3 Power combiner

The power combiner is a passive device used to combine two signals to a common output signal. The basic theory of power couplers will not be explained here, this can be found in [38]. The only major property that shall be mentioned here is the fact, that a lossless reciprocal three-port power combiner can not be matched for all inputs [38]. For this work, the simplest power combiner, the T-junction is used. It comprises three ports, two input and one output ports, all connected in a common point. The purpose of the power combiner is to efficiently couple both, the FFO radiation as well as radiation coming from e.g. the antenna to a single transmission line, which is then transmitting the power towards the mixer. In order to achieve a high transmission from the two inputs, towards the output, the impedances of the transmission lines have to be selected carefully. A schematic of such a T-junction is depicted in figure 5.9a comprising two input ports P2 and P3 as well as an output port P1. Since not all ports can be matched, the T-junction has to be designed to the particular purpose. In the frame of this work, the most important quantity is the power transmission from the various signal sources to the mixer.

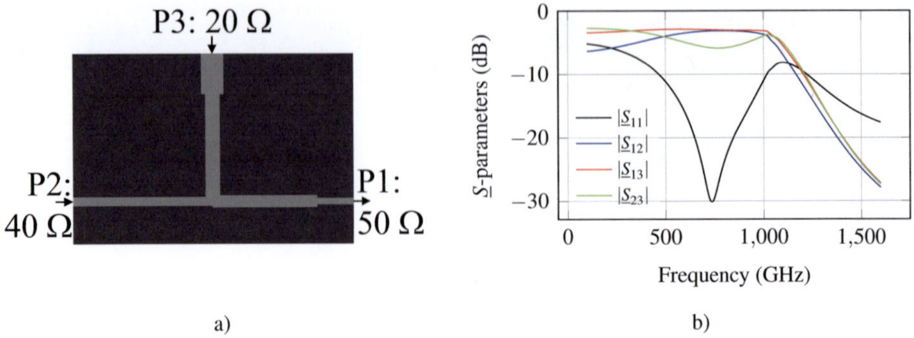

Figure 5.9: Exemplary design of a T-junction power combiner. This power combiner was designed for a maximized transmission at 750 GHz for the input and output impedances given in (a). The respective $\underline{S}$-parameters are given in (b). The transmission from the input is nearly ideal with a transmission of $|\underline{S}_{21}| = |\underline{S}_{31}| = -3$ dB. The output port 1 is designed to be matched which can be seen at the low reflection value $|\underline{S}_{11}|$.

Therefore, the output port of the T-junction was selected to be matched. To demonstrate the properties of such a T-junction, a simple example is discussed with a transmission line towards the mixer exhibiting $Z_{\text{out}} = 25\ \Omega$ and a characteristic impedance of the input lines of $Z_{\text{in,i}} = 50\ \Omega$ each. Since the two input ports are connected in parallel, the resulting circuit of both transmission lines will exhibit an impedance of 25 Ω, matching the output port, which results in a low input return loss. However, the other ports are not perfectly matched, which is why the maximum transmission of the input ports to the output results in $S_{\text{out,in,max}} = -3$ dB. Furthermore, the inputs are not well isolated from each other, resulting in a signal transmission into the other input port. In the particular case, the matching of the input is limited to $S_{\text{in,in}} = -6$ dB, whereas the transmission between the input ports is also $S_{\text{in1,in2}} = -6$ dB. By changing the input impedances, a weighting of the particular inputs can be achieved, however, as of now, both inputs are to be treated equally. If the characteristic impedances of the various transmission lines do not match the values given in the example above, the additional $\lambda/4$ transformers are required, modifying the impedances to values $Z'_{\text{in,i}}$ and Z'_{out}, which obey

$$\frac{Z'_{\text{in,1}} \cdot Z'_{\text{in,2}}}{Z'_{\text{in,1}} + Z'_{\text{in,2}}} = Z'_{\text{out}}. \tag{5.1}$$

As an example, input impedances of $Z_2 = 40\ \Omega$, $Z_3 = 20\ \Omega$ and an output impedance of $Z_1 = 50\ \Omega$ are assumed. The input impedances can, where necessary, be transformed to $Z'_{\text{in,i}} = 40\ \Omega$ resulting in $Z'_{\text{out}} = 20\ \Omega$. This can be transformed by a $\approx 32\ \Omega$ quarter-wavelength transformer to the output impedance of $Z_{\text{out}} = 50\ \Omega$. Figure 5.9a depicts the schematic of such a T-junction, with the impedances given in the example above. Each port of the T-junction contains a $\lambda/4$-transformer, before all three ports are connected in the center. Due to the non-ideal connection in the center of the structure, the exact values for the line lengths and widths need to be found by simulation. The simulation results of the optimized structure are shown in figure 5.9b. The return loss of the mixer port $|\underline{S}_{11}|$ exhibits a reflection as low as -30 dB at the design frequency of 750 GHz. The transmission exhibits approximately -3 dB from each input to the output port. The isolation is at -6 dB. Since all ports of the T-junction are conventional transmission lines, the waveguide port is used to excite each of them in MWS.

5.2.4 Antenna

In the frame of this work, double-dipole as well as double-slit antennas were realized. The double-dipole antenna was realized only once, since it proved to be more challenging to fabricate. A design example thereof is given in 5.10a for a design frequency of 166 GHz. In the left side of the schematic, the entire antenna is visible, where the dipoles are in the ground electrode as well as in the wiring layer. Each pole of a dipole is in different metallization layers. In figure 5.10a, the top half of the dipole is in the wiring layer, whereas the lower half is in the ground electrode. Each pole extends to the center, where it is connected to its respective metalization layer. The dipoles are connected by a 1 µm wide microstrip line. In order to characterize the antenna in the simulation, typically the $|\underline{S}_{11}|$ parameters are monitored.

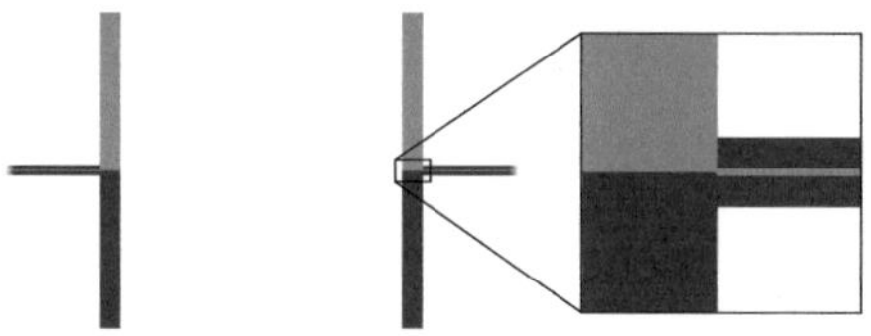

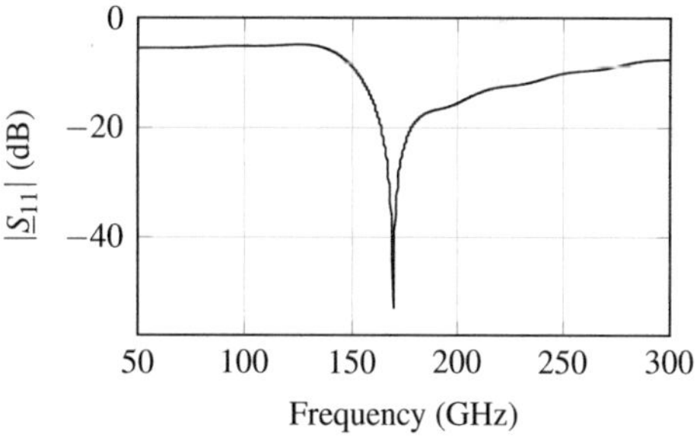

Figure 5.10: (a) Design of a double dipole antenna. The two dipoles seen in the design on the left are realized with one half in the ground electrode and another half in the wiring layer each. The microstrip line connecting the individual antenna arms is depicted in the close-up as well in (a). This antenna was designed for an operation frequency of 166 GHz. The corresponding reflection from the antenna can be seen in (b).

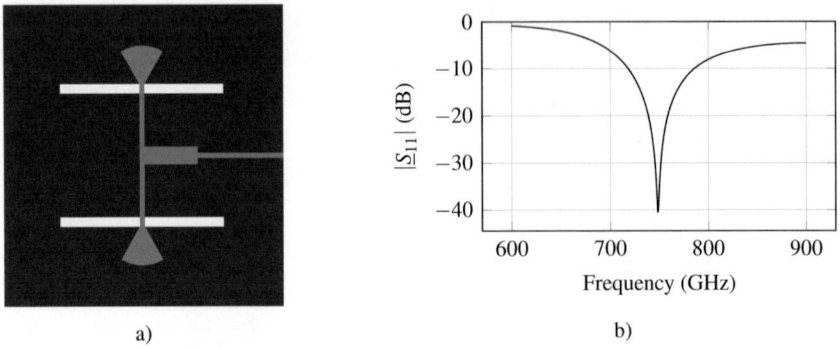

Figure 5.11: (a) Design of a double slot antenna designed for an operation frequency of 750 GHz. The two white rectangles depict the slots in the ground electrode (dark gray), which are transversed by stub-terminated microstrip transmission lines (light gray). The microstrip lines join in the center of the image where the transmission line network is connected through a $\lambda/4$ transformation. The corresponding $|\underline{S}_{11}|$ simulation results are depicted in (b), showing an excellent matching to the 40 Ω transmission line output.

For a matched antenna, the $|\underline{S}_{11}|$ parameter exhibits low values, which shows a good impedance matching between the antenna and the microstrip connection. Figure 5.10b depicts the resulting $|\underline{S}_{11}|$ simulation of this antenna. The simulation shows a good matching at the design frequency with a reflection of ≈ -50 dB. The space consumption of the double-dipole antenna in this simple configuration is however rather large. This is obvious, if the connectors to the antenna are considered. As shown in figure 5.10a, the connectors point outwards from the antenna. In order to feed both dipoles to the mixer, these microstrip lines meet at a certain point on chip. This however results in a rather large space consumption of the double-dipole antenna as compared to the double slit antenna, which is depicted schematically in figure 5.11a. Both slits are realized as trenches in the ground electrode which get filled by the planarization layer. The connection to the antenna is done by a microstrip line that is short circuited right outside of the trenches by radial stubs. The microstrip lines merge in the center of the antenna, and after a $\lambda/4$ transformation, the 40 Ω microstrip line connects the antenna to the transmission line network. Since the connectors of the antenna simply reach to the inside and do not have to circle around the entire antenna, the space consumption on chip is less. The performance of both antenna types in conjunction with a lens has been demonstrated to be almost identical [54, 55]. Therefore, most of the designs shown in later sections are realized using the double-slot antenna. Figure 5.11b shows the simulation result for this antenna. The design frequency of this design was 750 GHz. The matching exhibits values down to −40 dB.

5.2.5 Mixer, tuning and IF out

After the FFO signal is combined with the antenna signal in the T-junction, the signals are both fed to the mixer. In order to transmit the maximum amount of power, the mixer needs to be matched to its surrounding. This matching is the most crucial part of the entire design. It may include a matching of the real part of the mixer and always includes the tuning of the junction capacity. The necessity of compensation of this imaginary part was demonstrated in section 3.2.2, where this tuning was shown in an abstract manner. The inductive element can be realized using a short transmission line terminated with a short circuit. The short can not be realized as a physical connection between the wiring layer and the ground electrode, since it is in parallel to the mixer, which would prohibit the DC-biasing of the mixer. Therefore, a RF-short is used, which can be realized using an open-ended $\lambda/4$ stub. Due to the limited bandwidth of the $\lambda/4$ transformation, in literature almost exclusively, radial stubs are used. For low frequencies, these stubs have proven to exhibit excellent performance [92]. However, due the fact, that the stubs are not a physical short, a residual resistance remains, which gains in importance for higher frequencies. Figure 5.12a schematically depicts the top view of a mixer tuned with a transmission line terminated with a radial stub as a short. The mixer is indicated by the red square. The distance between the mixer and the tip of the radial stub is called the tuning length $\ell_{\mathrm{MS,tune}}$, since this is the length of the inductive transmission line.

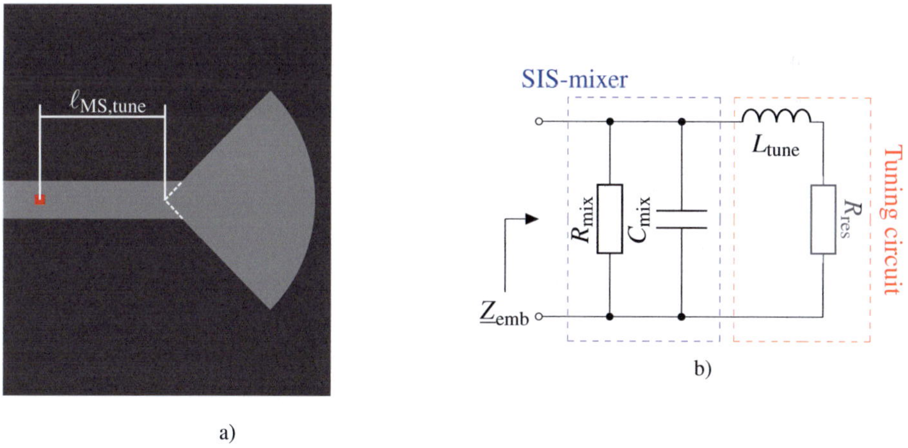

Figure 5.12: (a) Design of a single stub tuning of a SIS mixer. The mixer is indicated by the red square in the left part of the image. A microstrip line connects the mixer to the transmission line network (towards the left) and to a stub-terminated piece of microstrip line towards the right. The equivalent circuit of this design is given in (b).

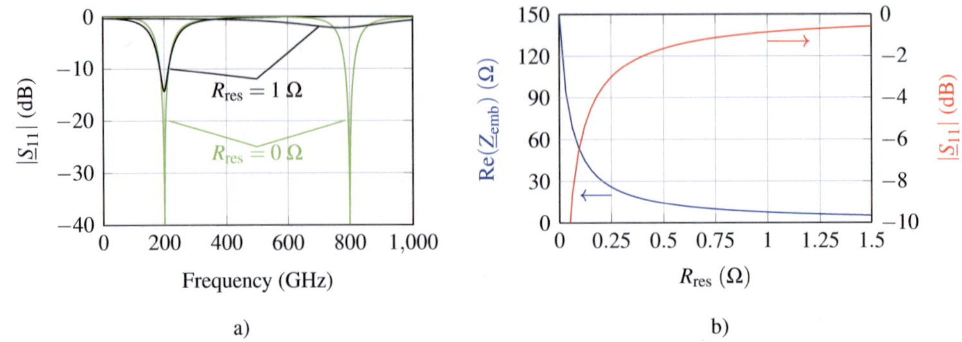

Figure 5.13: Simulated influence of the serial resistance in the mixer tuning circuit. The reflection parameters of a mixer with $C_{mix} = 76$ fF and a single stub tuning circuit are given in (a) for different frequencies. The different frequencies are achieved by the selection of the inductance to $L_{tune} = 8.33$ pH: $\nu = 200$ GHz and $L_{tune} = 0.52$ pH: $\nu = 800$ GHz. The green lines depict the resulting reflection in the ideal case, i.e. $R_{res} = 0$. A finite resistance $R_{res} = 1\ \Omega$ deteriorates the matching. The deterioration of the reflectance is due to the strong decrease in input resistance of the circuit as depicted in (b) along with the reflection for a frequency of 750 GHz.

The equivalent circuit of this structure with a non-perfect stub with a residual resistance (R_{res}) is depicted in 5.12b. Note, that this equivalent circuit is only valid for the design frequency. However, these circuits are used here and in the following to explain the RF behavior especially at the design frequency. Therefore, they do not contain the actual components, but display their behavior by the means of their respective equivalent circuit. By setting the length of the microstrip line $\ell_{MS,tune}$, the susceptance parallel to the mixer is adjusted, compensating the mixer reactance for a certain frequency. The connection of the inductor in parallel to the capacitor is realized by the short circuit of at the end of the transmission line. To demonstrate the effect of the finite parasitic resistance of the radial stub, the calculated $|\underline{S}_{11}|$ parameters are shown in figure 5.13a for two scenarios. The capacity is in both cases assumed with $C_{mix} = 76$ fF, the inductance exhibits two values $L_{tune,1} = 8.33$ pH and $L_{tune,2} = 0.52$ pH. These values result in resonance frequencies of $\nu_{res,1} = 200$ GHz and $\nu_{res,2} = 800$ GHz respectively. The green curves depict the resulting reflection calculation for an ideal short. The matching at the resonance frequencies is of course ideal. However, if the series resistance is increased, the effect on these curves is significant. An inserted series resistance of $R_{res} = 1\ \Omega$ results in the black curves. Whereas the circuit resonating at 200 GHz still exhibits a reflection of −10 dB, the 800 GHz resonator is already at $|\underline{S}_{11}| > -3$ dB, thus reflecting more than 50 % of the incident power.

These calculations are done, using the Spice feature of MWS from the equivalent circuit, without the inclusion of transmission line losses. However, compared with the residual resistance of the radial stub, the Ohmic losses on the short transmission lines are negligible. The

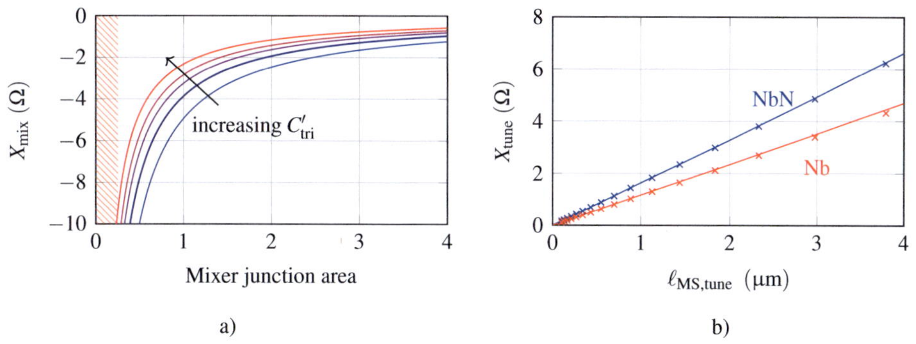

Figure 5.14: (a) The calculated reactances created by the mixer capacity at an operation frequency of $\nu = 700\,\text{GHz}$ for different trilayer capacitances ($C'_{tri} = 46\,\text{fF}/\mu\text{m}^2, 59\,\text{fF}/\mu\text{m}^2, 70\,\text{fF}/\mu\text{m}^2, 80\,\text{fF}/\mu\text{m}^2, 97\,\text{fF}/\mu\text{m}^2$). The patterning limit of the junction area is indicated by the red hatched area. The required tuning lengths for the particular capacitances can be read from (b). The resulting reactance depending on the tuning length are given for niobium as well as niobium nitride technology.

value of acceptable residual resistance can be derived from figure 5.13b. The blue line depicts the embedding impedance resulting from an increasing residual resistance. The strong decay of the embedding impedance is responsible for the increase in reflection, which is depicted with the red line. Therefore, the careful optimization of the radial stub with respect to a minimum residual resistance is of utmost importance.

Apart from the radial stub, the mixer tuning requires the inductive compensation part, which is realized using a transmission line. The required reactance is determined by the targeted operation frequency ν_{res} and the mixer capacity. The mixer capacity can be determined using the trilayer capacitance and the junction area. Figure 5.14a depicts the resulting reactances for $\nu_{res} = 700\,\text{GHz}$ depending on the junction area for different capacitances. For the NbN/AlN/NbN technology, this corresponds to critical current densities of $4, 6, 8, 10, 14\,\text{kA}/\text{cm}^2$. Clearly visible is the decreased X_{mix} for higher critical current densities and larger junction areas. The red hatched area reaching from $A = 0$ to $A = 0.25\,\mu\text{m}^2$ results from the minimum junction size of $0.5 \times 0.5\,\mu\text{m}^2$ and thus the patterning limit of the junctions. The values for the reactance obtained by field simulations of a shorted microstrip line are depicted in figure 5.14b. This simulations were conducted based on the material parameters for the NbN-based as well as for the Nb-based technology. Due to the shorter wavelength, the NbN-based microstrip line exhibit higher reactances than the Nb-based microstrip lines for a given length. The characteristic impedance for both transmission lines was $Z_0 = 30\,\Omega$ in order to compare the technologies to each other. The solid lines depict the fit with equation (2.52) and the wavelengths of $\lambda_{NbN} = 116\,\mu\text{m}$ and $\lambda_{Nb} = 162\,\mu\text{m}$.

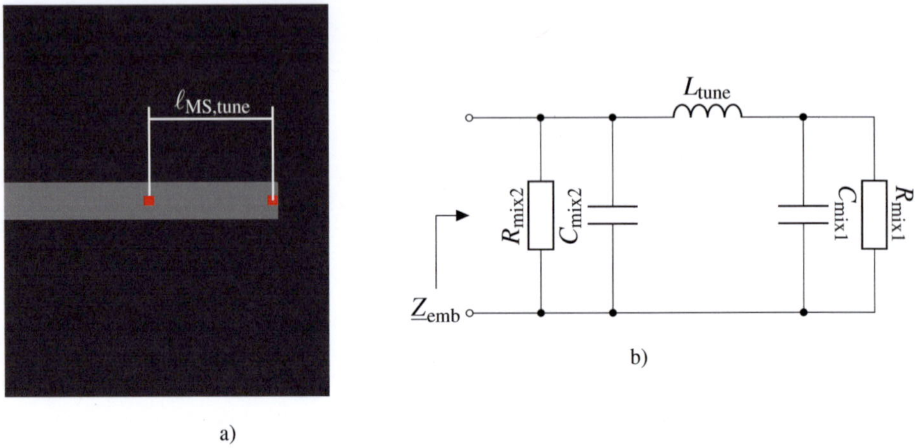

Figure 5.15: (a) Design of a twin-junction tuning circuit. The mixers are indicated by the red squares and are separated by the tuning length $\ell_{MS,tune}$. The corresponding equivalent circuit is depicted in (b).

In order to compensate the mixer reactance, the microstrip must exhibit

$$X_{tune} = \omega L_{tune} = -X_{mix} = \frac{1}{\omega C_{mix}}. \tag{5.2}$$

For the 30 Ω NbN transmission line, a reactance smaller than $X \leq 1.8\,\Omega$ can be realized only by sub-micrometer long transmission lines. Therefore, the lateral patterning limits are of utmost importance. For the demonstrated 700 GHz simulations, the compensation by a microstrip line is still possible with NbN, however, if the frequency increases, the $|X_{mix}|$ value further decreases, which can not be compensated by the transmission line transformation, since the tuning length becomes too short.

Other ways of tuning have been established, in order to overcome this challenge. A tuning method frequently used is the so-called twin-junction (twinJJ) tuning [92, 144], which also employs a transmission line transformation to compensate the mixer reactance and a second junction in a distance of the tuning length $\ell_{MS,tune}$. The schematic top view of such a structure is depicted in figure 5.15 along with the equivalent circuit. Since the inductor compensates for both capacities, larger inductance values are required as compared to the classical stub-tuning. Assuming identical junctions, the design rule for the compensation is

$$\omega L = 2\frac{1}{\omega C_{mix}}. \tag{5.3}$$

This larger inductance is beneficial especially for higher frequencies, since the transmission

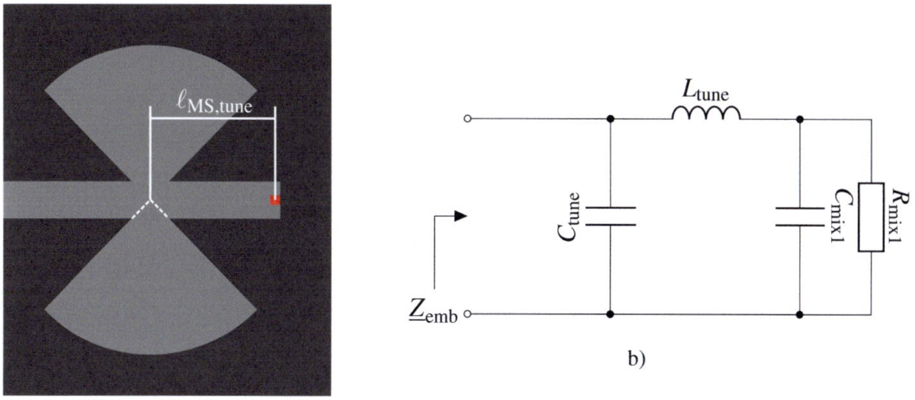

a)

Figure 5.16: Design of the hybrid end-loaded stub tuning approach is depicted in (a). The mixer is connected to an inductive stub which connects to a butterfly stub, providing a capacity in the distance $\ell_{\mathrm{MS,tune}}$. The equivalent circuit is depicted in (b).

lines are longer. The bandwidth of this tuning is of the same order as the classical stub tuning described above. However, due to the second junction, other effects are introduced. The resulting resistance of the device is $R_{\mathrm{total}} = \frac{R_{\mathrm{N}}}{2}$, where R_{N} refers to the single junction resistance. The second junction also results in a larger required oscillator power. Since both junctions are connected to each other in both, the wiring layer as well as the ground electrode, this setup realizes a superconducting quantum interference device (SQUID). Such a device is typically used to measure magnetic fields, and as such, it is very sensitive to magnetic fields. Therefore, in this approach, issues with noise and instabilities arise. However, this approach is frequently used, when high operation frequencies are targeted, since the stub-tuning approach can not be applied.

Another approach, that sets out to combine the advantages of both previously described approaches is the hybrid end-loaded stub (h-ELS) approach. This approach is based on the twinJJ idea, however, in order to eliminate the disadvantages introduced by the SQUID, the second junction is substituted with a capacitor. This approach is depicted in figure 5.16. The capacitor can be realized by simply using a straight stub, however, using a radial stub, offers a wider bandwidth. The advantage of this design is the omission of the second junction, and thus the SQUID setup. Alike the twinJJ approach, higher frequencies are achievable as compared to the classical stub tuning, due to the longer tuning transmission line. However, the bandwidth of this setup is smaller, due to the larger capacitance. In the twinJJ design, the larger capacitance is compensated by the reduction of the resistance of the circuit (compare equation (3.56)). However, in the h-ELS design, the second resistor is not present, thus the higher capacity leads to a decreased bandwidth compared to the other design approaches.

RF bandwidth All three tuning approaches can be used for the compensation of the mixer capacity for a certain frequency. As given in equation (3.56), the RF bandwidth is determined by the critical current density of the trilayer in conjunction with the lateral dimensions of the junction. For a given j_c and area A of the junction, the achievable RF bandwidth of the different tuning approaches differ. The bandwidth achievable by the classical stub tuning can simply be calculated using equation (3.55). The twinJJ tuning however exhibits twice the capacity due to the parallel junctions. This on the other hand decreases the normal state resistance of the mixer circuit by a factor of two for identical junction, resulting in an identical bandwidth as compared to the classical stub tuning. By choosing slightly different junction areas, the bandwidth of the mixer can even be tuned to higher values. The hybrid end-loaded stub exhibits the same capacity as the twinJJ design, due to the parallel radial stub. However, due to the lack of the parallel resistor, the bandwidth of this tuning approach exhibits a lower bandwidth. In the case of a radial stub that exhibits the identical capacity as the mixer junction, the bandwidth is reduced by a factor of two as compared to the classical tuning approach or the twinJJ approach.

Intermediate frequency output The intermediate frequency output needs to be attached to the microwave circuit without disturbing the properties of the circuit. Therefore, it can be designed to behave as a high-Ohmic connection for the operation frequency of the receiver. Since the intermediate frequency is typically at significantly lower frequency levels, this high-Ohmic connection is not visible for the intermediate frequency. The radial stub as an example provides a short-circuit for the target frequency, however the intermediate frequency can pass the radial stub with negligible reflection. The intermediate frequency can therefore be attached to the outer edge of a radial stub, however, care has to be taken not to deteriorate the radial stub properties. Another option is the attachment of a quarter wavelength transmission line in extension to the tuning length, terminated by a radial stub. The short circuit provided by the radial stub will be transformed by the quarter wavelength transformer into an open circuit, resulting in the required high-Ohmic connection. This intermediate frequency connection thus has to be designed for each tuning approach separately, in order to match the particular geometrical requirements.

Control lines The mixer capacity compensation can be realized with the approaches introduced above. In order to operate the mixer in the desired quasiparticle regime, the Josephson

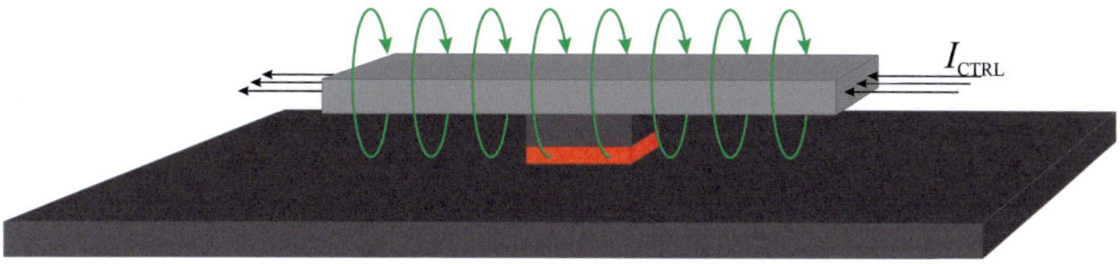

Figure 5.17: Schematic setup of a mixer control line realized in the wiring layer. The wiring layer on top of the mixer junction provides the mixer with the RF signal from the transmission line network. It can furthermore be utilized for the provision of the magnetic field by passing a DC current along the junction as indicated in the figure.

current needs to be suppressed. This is done by applying a magnetic field, which can completely suppress the Josephson current, see figure 3.8. The magnetic field therefore has to be large enough to at least introduce one fluxon inside the junction. In practice, the mixer is desired to be operated at even higher magnetic fields [1]. Since the mixer junctions are as small as possible, i.e. with areas smaller than 1 μm^2, the magnetic field required for the Josephson current suppression is quite high. Due to the fact, that the mixer and the flux-flow oscillator require different magnetic field densities, the mixer magnetic field is, as opposed to single mixer structures also applied to the mixer with a control line. This control line can be realized in the ground electrode, or in the wiring layer. If the mixer control line is realized in the wiring layer, it has to blend in the microwave design, hence, the geometries are predefined. Figure 5.17 schematically depicts a side view of the mixer attached to the microstrip line, which in this case is used as the control line. The red area represents the insulating barrier, which is not depicted to scale. The control current is fed along the wiring layer passing by the mixer. This approach is very practical, since the microstrip line above the mixer is required anyway, so it can be used in almost any design. The only requirement is an extra DC-connector at a point in the transmission line network, in order to extract the DC-control line current. The injection can be done via the IF out connection. The current-inherent magnetic field surrounding the microstrip line penetrates the barrier, thus suppressing the Josephson current.

Due to the limited degrees of freedom in designing the control line in the wiring layer, it may be beneficial to realize the control line in the ground electrode. However, since this layer is also part of the microstrip line, the control line can not be realized within the area of the RF network. This can be overcome, if the intermediate frequency output is realized using a coplanar waveguide instead of continuing with the microstrip transmission line as depicted in figure 5.18.

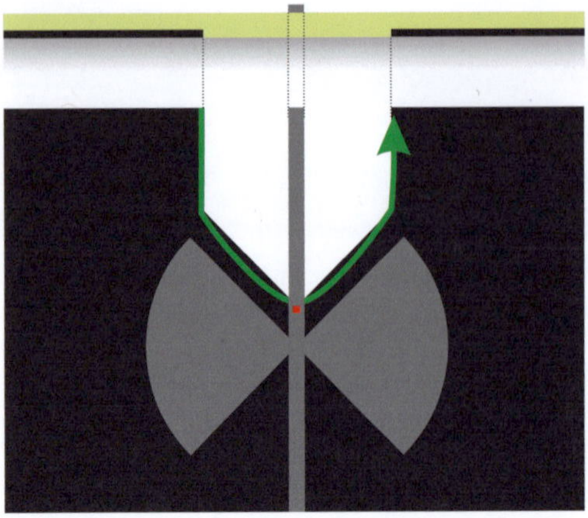

Figure 5.18: Schematic design of a mixer control line in the ground electrode. For the usage of a hybrid end-loaded stub tuning approach, the ground electrode after the mixer can be cut out, guiding a current along a path indicated by the green arrow. The IF output of the mixer in this case needs to be realized using a coplanar design as shown in the cross-section in the top part of the image.

Although, the center conductor is not in the same metalization layer as the ground plane, the structure very well behaves as a coplanar waveguide, since the vertical distances are negligible compared to the significantly larger lateral dimensions. Figure 5.18 also depicts a cross-section through such a coplanar waveguide. The mixer tuning should however not be affected by the connection of this IF output. Therefore, the IF out connection is realized using a stepped impedance filter [38]. This stepped impedance filter is designed in such a way that it provides the mixer with a very high resistance and a zero reactance, so neither the mixer tuning is affected, nor the total resistance of the mixer embedding impedance. Using this gap in the coplanar waveguide, a current can be driven through the ground electrode from the one end of the CPW ground to the other as indicated by the green arrow in figure 5.18. The magnetic field generated by this current can suppress the Josephson current of the mixer. This approach can obviously not be used for all tuning approaches. The most obvious usage is of course the hybrid end-loaded stub. In this case, the mixer is the last device in the RF-signal chain. In the classical stub tuning approach, the missing ground under the radial stub might lead to tuning complications. Since the current is carried in the wide ground electrode, the current flow is not homogeneous, but concentrated at the edges as indicated by the green arrow in figure 5.18. This effect is particularly exploited by this approach, focusing the current to the edges, which are in the vicinity of the mixer junction. In the twinJJ tuning approach, the ground electrode might also be used as control line. However, this was not tested within this work, which is why this approach is not discussed in detail.

Conclusion

In this section, the designs of the various components and their particular requirements are introduced. This includes the impedance matching of the FFO to the significantly higher impedance of the transmission line network by using n-stage quarter wavelength transformers. The separation of the FFO from the mixer for DC signals is achieved by the usage of DC breaks, which can be realized in both, the ground electrode as well as the wiring layer. The design of two types of antennas is introduced, the double dipole and the double slot antenna. The FFO signal and the antenna signal are combined in a T-junction, which can realize a -3 dB transmission of each of the input signals to the output. The most challenging part is the tuning of the mixer capacity, especially for high frequencies. Three approaches of this tuning are discussed, where each approach is based on the transmission line impedance transformation. Therefore, the transmission line properties are of utmost importance. Due to the high kinetic inductance, the microstrip lines responsible for the tuning become extremely short for high frequencies, and hence, the classical tuning approach reaches a limit where the tuning length is deep in the sub-micrometer range. This is very unpractical and therefore, other approaches, such as the twinJJ tuning and the hybrid end-loaded stub tuning are introduced as well. The final component introduced in this chapter is the mixer control line. This is typically realized in the wiring layer, however, the geometry of the control line is in this case dictated by the RF properties of the transmission line which connects the mixer to the periphery on chip. A different approach, which was developed during this work, is the realization of the mixer control line in the ground electrode. This however inhibits the usage of the classical stub tuning. This approach can be realized with the subsequent components in a coplanar technology.

Designing each of this parts is done, considering the material parameters of the particular fabrication technology. In the schematic tool, these components can then be connected and simulated as an entity. This is the basis for the development of the circuits presented in the following chapters.

6 Prototype devices with reduced complexity

The technology development and the simulations have been discussed in the previous chapters. In order to realize a fully integrated receiver, the material parameters and the simulations need to be verified. Due to the high frequencies, the single components can not be characterized towards their RF properties separately. Since the fully integrated receiver is a complex system consisting of many different components, preliminary devices were realized, in order to split the investigation into smaller, less complex systems. Apart from the reduced complexity, especially when leaving out the antenna in order to investigate on the coupling of the FFO to the mixer, the space consumption on a chip per device is decreased dramatically, which enables the realization of many of those prototypes on a $10 \times 10\ \text{mm}^2$ sized chip, allowing for parameter sweeps to systematically investigate the various prototypes.

The different devices have been fabricated at intermediate technology development levels during this work. Therefore, the prototypes are realized in both, pure Nb technology as well as NbN technology. The mature niobium technology is realized on oxidized silicon substrates, whereas the niobium nitride based prototypes are realized on both, oxidized silicon as well as sapphire substrates.

6.1 Mixer response to an external 166 GHz source

The classical mixer without a local oscillator on chip is in this case the prototype with the least complex structure. The preliminary device consists of a double-dipole antenna, which is coupled to a mixer by a microstrip line. The device was designed for a frequency of 166 GHz, expecting the local oscillator signal to be provided by an external source. The devices were realized in NbN/AlN/NbN technology on oxidized silicon substrates.

A quasi-optical characterization can not be done in a liquid helium transport dewar. Therefore, a setup using a bath cryostat was constructed, allowing the characterization of the mixer response to an external 166 GHz source. This setup is shown schematically in figure 6.1.

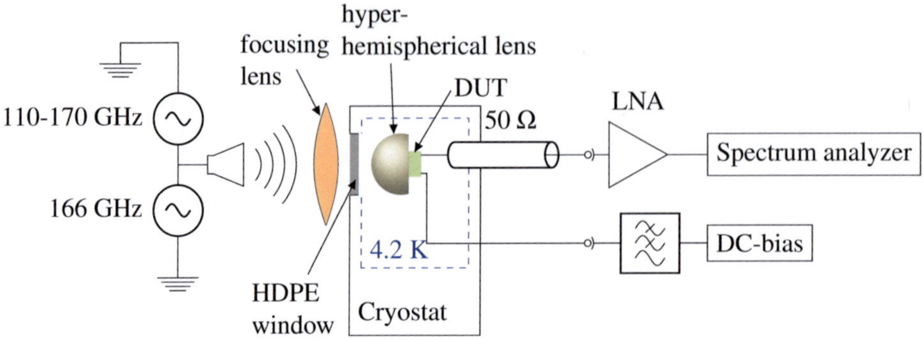

Figure 6.1: Optical setup for the measurement of the mixer response to external radiation. Both quasioptical signals are generated by an active frequency multiplier source. A 166 GHz source and a low power tunable source are connected to a waveguide which is terminated with a horn antenna, facing the cryostat. The cryostat contains the mixer installed on a hyperhemispherical lens inside a detector block. The intermediate frequency output of the mixer is fed through a 50 Ω coaxial transmission line to a room-temperature amplifier. The output signal can then be recorded by a spectrum analyzer.

The mixer chip is mounted inside a liquid-helium bath cryostat, which provides access to the external radiation through a high-density polyethylene (HDPE) window. The local oscillator signal is generated by source with a fixed frequency of 166 GHz which exhibits a high power level of 11.4 mW. The signal of interest is generated by another source, which can generate frequencies between $110 - 170$ GHz with lower power levels of ≈ 208 μW. Both sources are waveguide coupled and connected to each other with a waveguide power combiner that is terminated with a horn antenna, enabling the transition from the waveguide to free-space optics. The signal is then focused by a first lens onto a hyperhemispherical lens inside the cryostat. On the backside of this hyperhemispherical lens, the mixer chip is mounted, embedded in a connector printed circuit board. This connector board provides the necessary connections to the mixer chip, i.e. the bias connections as well as a coplanar waveguide for the intermediate frequency output. The lens with the mixer and the connector board are all mounted inside a copper detector block, which is equipped with DC plugs for the biasing and SMA-connectors for the intermediate frequency output, which is then fed through a coaxial 50 Ω transmission line to a room-temperature low noise amplifier (LNA). The output of the LNA can then directly be displayed on a spectrum analyzer. The DC-bias is provided through a filtered twisted-pair set of copper wires.

The equivalent circuit of the realized mixer is depicted in figure 6.2. The antenna in the very left is designed to exhibit an output impedance of 30 Ω, from which a microstrip transmission line transfers the signal to the mixer. The mixer junction is tuned using the classical

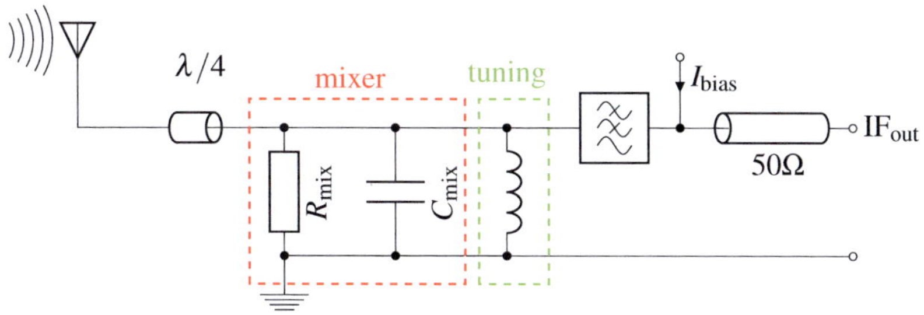

Figure 6.2: Equivalent circuit of the 166 GHz mixer design. The signal is received by a double-dipole antenna and then transmitted to the SIS mixer. The tuning is achieved using classical stub tuning.

stub tuning, since for these low frequencies, the disadvantages of this tuning approach are not relevant.

A SEM image of the realized structure can be seen in figure 6.3. The two antennas can be recognized in the center of the image, realized as $\lambda/2$ dipoles with dimensions of $\ell = 305$ µm and $w = 20$ µm.

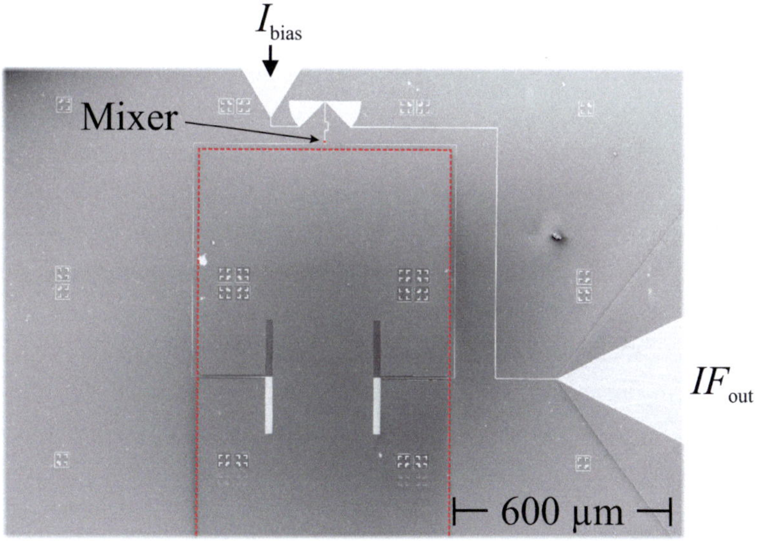

Figure 6.3: SEM image of a realized 166 GHz mixer. In the center of the image, the antenna can be recognized. The bright areas are the wiring layer, the darker areas are the lower electrode covered with SiO. The antenna is located in the center of a rectangular opening in the ground electrode which is illustrated by the dashed red lines. The antenna signal is transmitted to the mixer which is located in the top part of the image. Above the mixer, the butterfly stub can be recognized. At the ends of the butterfly stub, the DC bias line and the IF output microstrip line are connected. The IF microstrip line leads to a tapered coplanar waveguide. The onset of this taper can be seen in the right part of the image.

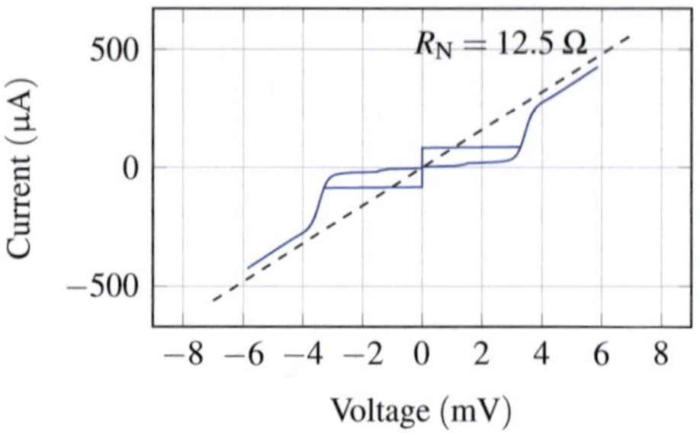

Figure 6.4: Current-voltage characteristic of a $0.9 \times 0.9\,\mu m^2$ sized mixer junction along with the design resistance, indicated by the black dashed line. Data adopted from [Mar16].

In this geometry, one dipole exhibits a characteristic impedance of $30\,\Omega$. In order to realize the dipole structure, one half of the dipoles is realized in the ground electrode whereas the other is realized in the wiring layer. This is the reason, why the different parts of the antenna appear in different colors in the SEM picture. The pole connected to the ground electrode (top part in the image) is covered by SiO, and therefore appears in a darker gray as compared to the pole realized in the wiring layer (bottom half in the image). The antenna is located in a rectangular opening in the ground electrode. The opening can be recognized in the center of the image. Surrounding the top half of this opening, the microstrip line is located, merging the two antenna arms in a T-junction. The microstrip line then feeds the signal to the mixer junction which is indicated by the red square in the top part of the image. Different junction resistances were targeted, where the impedance transformation was realized with a $\lambda/4$ transformation. The reactance tuning is done by a compensation microstrip line which is terminated with a butterfly stub that can be recognized in the very top part of the image. At one side of the butterfly stub, the DC connection is realized. The microstrip line emerging from the right end of the butterfly stub is designed as a $50\,\Omega$ microstrip line, which provides access for the intermediate frequency output. Connecting this to the CPW on the connector board requires the microstrip line on chip to be transformed to a coplanar waveguide, which can be seen in the right end of the SEM image. The CPW line is then tapered to match the dimensions of the CPW on the connector board.

A set of 9 different designs was realized on this chip with varying mixer resistances. As an example, figure 6.4 depicts the current-voltage measurement of a junction, designed to exhibit a resistance of $12.5\,\Omega$. With a measured value of $R_N = 12.43\,\Omega$, the normal state resistance of the junction is very close to the designed value. This junction is realized with lateral dimensions of $1.2 \times 1.2\,\mu m^2$ and a critical current density of $j_c = 7\,kA/cm^2$. The expected characteristic values as shown in chapter 4 are obtained. The mixer devices are

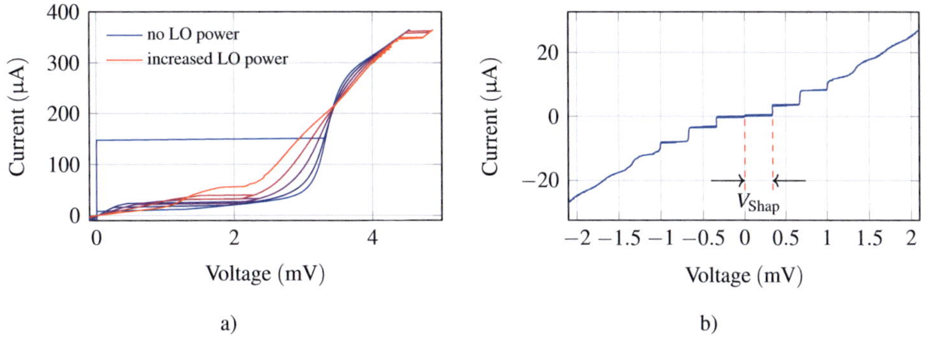

Figure 6.5: Optical response of the 166 GHz mixer to the local oscillator signal. (a) Set of mixer current-voltage characteristics with varied LO power. The multiple curves correspond to different incident local oscillator powers. The magnified image in (b) depicts a close up of the mixer current-voltage characteristic close to $V_0 = 0$, showing the Shapiro steps up to the 4th order. The span of each Shapiro step is 0.343 mV. Data adopted from [Mar16].

first irradiated with the local oscillator signal. Recording the current-voltage characteristic for different LO power levels yields the mixer response to the local oscillator signal. The result of this investigation is depicted in figure 6.5a. The blue curve represents the junction characteristic without external radiation as a reference. Clearly visible is the emergence of the photon-assisted tunneling with increasing LO power. Increasing the power of the local oscillator results in a higher current in the quasiparticle branch at a given voltage. Due to the lack of a control line, the mixer exhibits Shapiro steps, which can be used to determine the frequency of the incoming radiation. An example of recorded Shapiro steps is depicted in figure 6.5b. Clearly visible is the step-like behavior starting from zero voltage with a step-width of 0.34 mV. This corresponds to a frequency of the incident radiation of 166 GHz (see equation 3.26). The Shapiro steps are visible up to the fourth order. The device hence behaves as expected, the local oscillator radiation is transmitted to the mixer junction, manifesting itself in the expected step like features in the current-voltage characteristic.

The previous investigations are all based on monitoring the current-voltage characteristic with LO radiation exclusively. However, for the mixer operation, the signal of interest needs to be applied as well and the IF output needs to be monitored. The voltage of the mixer is therefore set to a constant value on the first photon step, e.g. 3.5 mV and the output signal through the intermediate frequency output is monitored using a spectrum analyzer. The employed amplifier in this case was a room-temperature low-noise amplifier. For best noise properties, this amplifier would need to be located at the 4.2 K stage. The applied signals are depicted in figure 6.6a. The spectra shown are reconstructed from the measurements of the spectrum analyzer, which are depicted in figure 6.6b.

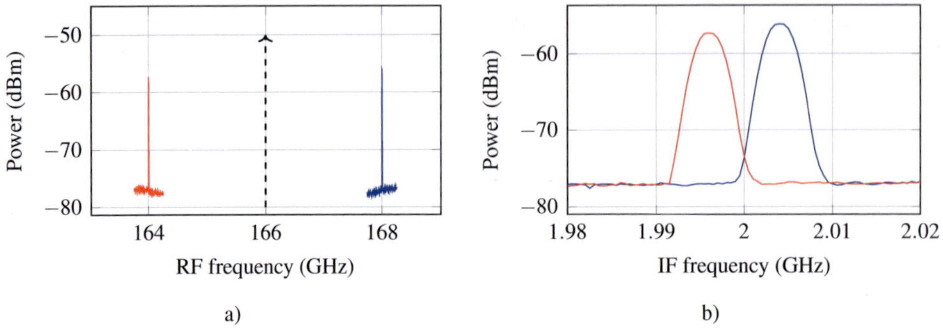

Figure 6.6: (a) Excitation of the mixer with 164 GHz representing the lower sideband and 168 GHz for the upper sideband. This is depicted in along with the LO frequency of 166 GHz. (b) The down converted signals are depicted at ≈ 2 GHz. Data adopted from [Mar16].

Both, the upper as well as the lower side band have been measured. The local oscillator power was in this case set to approximately 1 mW. The incident signal is first set to a frequency of 168 GHz, the result is depicted in figure 6.6 which is in this case the upper side band. The red curve depicts the measurement for the incident signal being set to a frequency of 164 GHz in figure 6.6a. The black dashed arrow illustrates the local oscillator frequency. The down-converted spectrum is depicted in figure 6.6b. Both, the upper and the lower side band signal were set at a frequency 2 GHz apart from the center frequency of the local oscillator signal. Therefore both signals should be downconverted to a intermediate frequency of 2 GHz. However, due to a small constant frequency offset of the source of 4 MHz, the down-converted spectra differ by a frequency of 8 MHz. Even such small frequency differences of only 4 MHz can be easily resolved. The slight difference in the IF output power can in this case be explained by variations in the available output power of the signal source, which is not perfectly constant across the entire spectrum.

Determining the spectral behavior of the device can be done by variation of the RF signal frequency. The tuneable source is specified to cover a frequency span from 110 to 170 GHz. Since the high-power source has a fixed frequency of 166 GHz, the frequency behavior can be investigated in a wider frequency range when operating the mixer using the lower side band. The result of the response from the mixer on the varied frequency of the RF signal is depicted in figure 6.7. The blue line depicts the envelope of a sweep in RF frequency. The black curve depicts the measured noise level, without any incident RF power. Clearly visible is the inclination of the measured IF power starting from 0 GHz up to approximately 3 GHz and a strong decrease for frequencies higher than 8 GHz. This is explained by the bandwidth of the amplifier. For the maximum tuning range of the source, the IF signal

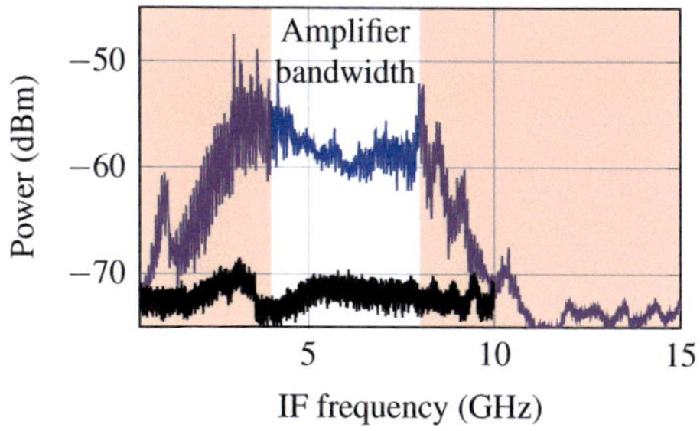

Figure 6.7: Measurement of the intermediate frequency output spectrum. The noise floor is depicted in black, the measured signal with applied radiation is depicted in blue. The bandwidth is limited by the amplifier, which has a specified transmission band reaching from 3 to 8 GHz as indicated by the white area in the spectrum. Data adopted from [Mar16].

should exhibit frequencies up to 56 GHz. However, the bandwidth of the employed amplifier is only specified from 4 up to 8 GHz as indicated by the white area in figure 6.7. Since the local oscillator frequency can not be changed, only frequencies between 162 and 158 GHz can be investigated. As shown in figure 6.7, the bandwidth limit of the amplifier becomes apparent for frequencies lower than 158 GHz, since this results in intermediate frequencies larger than 8 GHz. Therefore, the frequency range of the device can not be determined using this setup, since the component on chip with the smallest bandwidth still exhibits a bandwidth exceeding 16 GHz.

However, another noteworthy property of the device became apparent within this investigation, which concerns the power levels of the local oscillator and the RF-signal. Both sources contain attenuators in their particular waveguides, which can be adjusted mechanically. These attenuators are however not calibrated, so it is not possible to set a certain power level without measuring the power for a given attenuator position. The spectrum shown in figure 6.7 was recorded at the minimum power level that can be achieved using these built-in attenuators. Determining the power levels was done using a power meter. This power meter can resolve power levels down to 10 μW (= −30 dBm). This level is however already reached at an intermediate position of the attenuator. Hence, the power level of the local oscillator arriving at the cryostat can only be determined to be smaller than 10 μW. This power level is of course not the one actually reaching the output port of the antenna, due to reflection or absorption losses. These losses are characterized using the coupling efficiency, which describes the amount of power transmitted from free-space to the output port of the antenna in relation to the incident radiation power. Other experiments have shown a coupling efficiency of typically only a few percent [145, 146]. Assuming a coupling efficiency of 10 %, the power level present at the antenna can be calculated to less than 1 μW. Since the power levels of the RF-signal are significantly lower than the ones of the local oscillator signal, no reasonable estimation of the incident power level was possible with this setup.

Conclusion

The presented test device includes an antenna connected to a mixer with a classical stub tuning for a target frequency of 166 GHz. The successful demonstration of the mixer properties were achieved by a powerful non-tunable local oscillator and a less powerful but tunable second radiation source. The intermediate frequency spectrum was successfully obtained using a setup which is not yet optimized for an ultra-low noise operation. Due to a limited amplifier bandwidth and the non-tunable local oscillator source, the bandwidth of the mixer could not yet be determined. However, this preliminary device enabled the mixing of a 166 GHz signal.

6.2 Integration of a FFO with a mixer junction

After the successful investigation of the mixer connected to an antenna, an increase in complexity of the total system is done. This is the incorporation of a flux-flow oscillator with a mixer junction. Since the material parameters of NbN strongly depend on the stoichiometry, the degree of uncertainty is higher as compared to e.g. an elemental superconductor such as niobium. In order to keep the uncertainty at low levels, the initial FFO prototype is realized using the niobium technology. Furthermore, no antenna is present in the following devices, only the coupling between FFO and mixer junction is investigated.

6.2.1 400 GHz radiation transmission from a flux-flow oscillator to a mixer junction

A more complex structure than the previously discussed prototype is the mixer integrated with a flux-flow oscillator, since it includes more components. All components have to be designed and optimized to match the particular requirements as described in the previous chapter. In addition, the particular challenge when integrating an FFO is its low output impedance and the separation of the DC allowing for different biasing of FFO and mixer. In this first FFO-integrated prototype, only the DC break in the wiring layer is present, the ground electrodes of FFO and mixer are connected. Since this device is realized using pure niobium-based trilayers, the values for the Nb/Al-AlO_x/Nb trilayers apply. The trilayers are deposited on a $10 \times 10\,mm^2$ oxidized silicon substrate. On a substrate with such lateral dimensions, multiple variations of a design can be realized. The equivalent circuit of the investigated

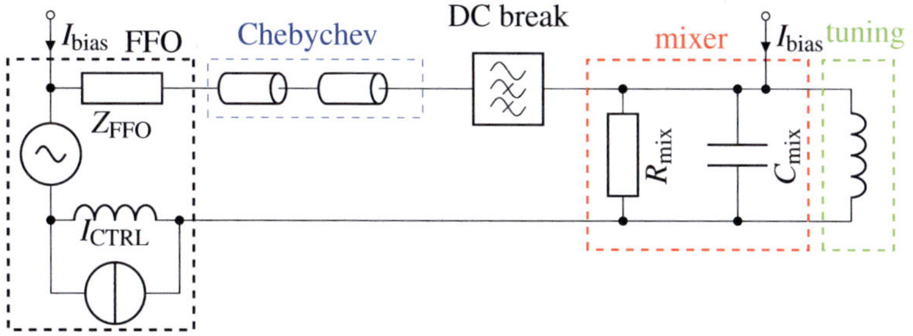

Figure 6.8: Equivalent circuit of the Nb-based 400 GHz test device, comprising a flux-flow oscillator connected to a SIS mixer. The FFO is depicted in the left part of the image, followed by a two-stage Chebychev matching transformer, a DC break in the wiring layer and the mixer, including the tuning circuit. Again, a classical stub tuning is used. The FFO control line is indicated by the inductor in the ground electrode.

structure is given in figure 6.8. The FFO, being in the beginning of the signal path, requires a control line for operation. This control line is included in the ground electrode. The output resistance of the FFO is depicted by the resistor and was calculated for this technology and geometry. Using equations (2.24) and (3.77)-(3.80) with the parameters $C'_{tri} = 85\ \mathrm{fF}/\mu\mathrm{m}^2$, $W_1 = 1\ \mu\mathrm{m}$, $W_2 = 6.5\ \mu\mathrm{m}$, $\lambda_L = 90$ nm, $t_1 = 90$ nm, $t_3 = 150$ nm $t_{diel} = 211$ nm, $h_1 = 215$ nm, $h_2 = 426$ nm, the FFO output resistance is determined to $Z_{FFO} = 0.64\ \Omega$.

This low impedance is then transformed using a Chebychev matching transformer to a characteristic impedance of 11 Ω. This impedance has been selected, since the following DC break can be realized with the microstrip width $w_{MS} = 4.5\ \mu\mathrm{m}$, which corresponds to this characteristic impedance. The FFO, the Chebychev transformer and the DC break are identical for all investigated structures. However, the mixer tuning was varied. A classical stub tuning was used in this case. In order to cover the various uncertainties, multiple designs for the tuning were created. The targeted design frequency range was selected to ≈ 400 GHz. These low frequencies in conjunction with the usage of pure niobium as transmission line material allows for the realization of the tuning using the classical stub tuning.

Figure 6.9 shows an SEM image of one example of the realized structures. The light gray areas are the wiring layer, the ground electrode appears darker, since it is covered with silicon monoxide. The flux-flow oscillator can be seen in the left part of the image, with the bias connections coming from top and bottom. The FFO is followed by the Chebychev transformer and the DC break. The last stage in this signal chain is the mixer junction with the tuning. The right part of figure 6.9 depicts a close-up of the mixer with the tuning. The tuning length is in this case $\ell_{MS,tune} = 9.25\ \mu\mathrm{m}$.

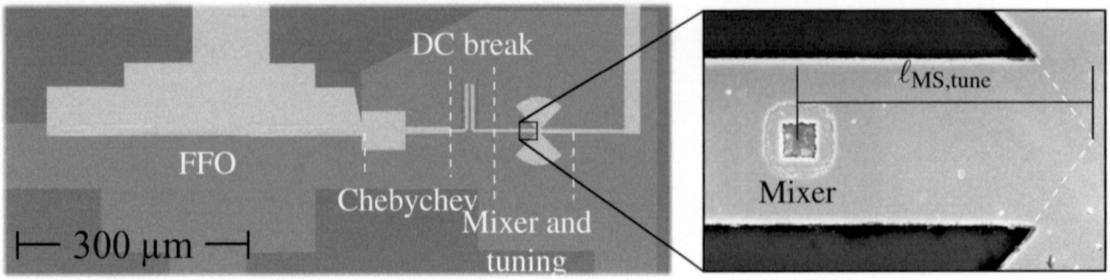

Figure 6.9: SEM image of one of the realized 400 GHz devices. The bias lines feeding the FFO are located in the left part of the image, pointing towards the top and bottom. The Chebychev matching transformer can be recognized at the FFO output, followed by the wiring layer DC break. The mixer is located right before the butterfly stub. A close-up of this region can be seen in the magnified image on the right.

The close up also reveals sharp edges in the connection point between the butterfly stub and the microstrip line, which approves the design accuracy of the patterning process. The mixer junction depicted in figure 6.9 exhibits lateral dimensions of $1.92 \times 1.92\ \mu m^2$ and is expected to exhibit a normal state resistance very close to the characteristic impedance of the microstrip line. Hence, an impedance transformation of the real part of the mixer impedance to the characteristic impedance of the transmission line is expendable. With the trilayer capacitance and this area, the mixer capacity can be calculated to $C_{mix} = 313\,fF$. Using equations (2.52) and (2.29) with the material parameters for this technology, the resonance frequency can be calculated by

$$\frac{1}{\omega C} = Z_0 \cdot \tan\left(\frac{2\pi \ell_{MS,tune}}{\lambda}\right) \tag{6.1}$$

for the structure given in figure 6.9 with $\ell_{MS,tune} = 9.25\ \mu m$ to

$$\nu_{res} = 282\ \text{GHz}. \tag{6.2}$$

Since there is no antenna present in this design, the characterization is done with the dipstick setup in the liquid helium dewar exclusively. An initial DC characterization of FFO and mixer separately reveals the functionality of the devices and allows the evaluation of the assumptions made prior to fabrication. Figure 6.10 depicts a series of measurements of the flux-flow oscillator. Each of these curves is recorded for a different control line current. After setting a control line current, the bias is increased from zero up to a value, where the FFO voltage jumps to the gap voltage. This set of curves is obtained for control line currents

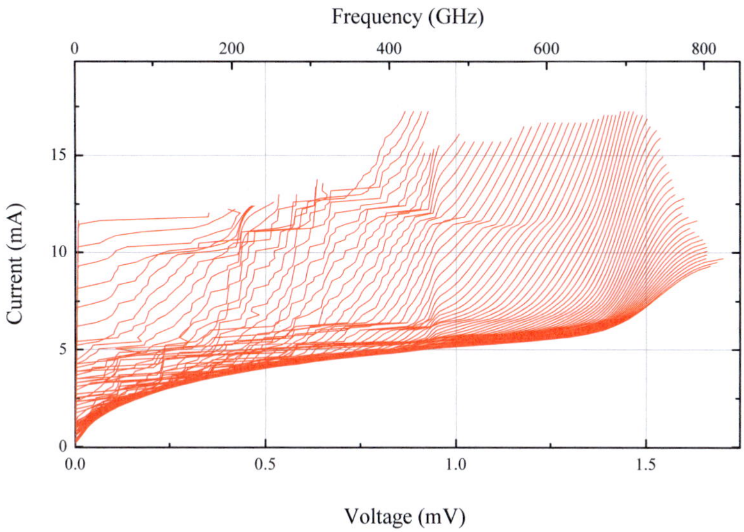

Figure 6.10: Set of current-voltage characteristics of one of the realized flux-flow oscillators. Clearly distinguishable is the Fiske regime from the flux-flow regime. The FFO can be operated up to the expected ≈ 700 GHz. Data adopted from [Kar13].

between $5\,\text{mA} \leq I_{\text{CTRL}} \leq 35\,\text{mA}$. The transition from the Fiske- to the flux-flow regime is visible at a FFO voltage of $V_{\text{FFO}} = 0.94\,\text{mV}$. The highest bias currents can be applied up to a maximum voltage of $V_{\text{FFO}} \approx 1.5\,\text{mV}$, corresponding to a frequency of $\approx 700\,\text{GHz}$, as expected for the niobium-based flux-flow oscillator.

After the observation of the FFO to behave as expected, the mixer is characterized. The DC current-voltage characteristic without any external influence is shown in figure 6.11 by the blue curve. The normal state resistance of this device (same as in figure 6.9), is determined to $R_{\text{N}} = 14.7\,\Omega$, which is close to the assumed value of $Z_0 = 11\,\Omega$. Setting the flux-flow oscillator bias point to a certain value, the influence on the mixer can be investigated by recording its current-voltage characteristic. Three example curves are depicted in figure 6.11. The mixer is subjected to three different FFO signals with different frequencies: $\nu_{\text{FFO}} = 338\,\text{GHz}, 357\,\text{GHz}$ and $386\,\text{GHz}$. The curves recorded under radiation significantly deviate from the original curve due to an overlay of Shapiro steps and photon-assisted tunneling steps. The frequency of the incident radiation is determined using the voltage of the flux-flow oscillator in the bias point. The Shapiro steps can be exploited to verify the radiation frequency, which indeed perfectly fit the frequency expected from the FFO voltage. The measured FFO voltages are depicted in figure 6.11 as vertical dotted lines and clearly match the voltages of the Shapiro steps.

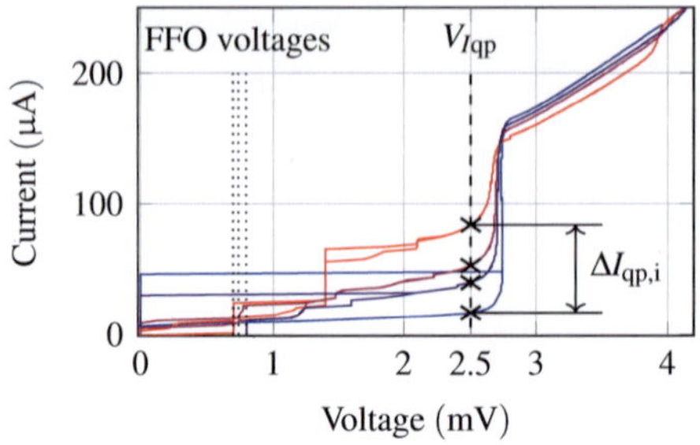

Figure 6.11: Current-voltage characteristics of a SIS mixer pumped by a flux-flow oscillator. The unpumped curve is depicted in blue, turning to red for an increased quasiparticle current rise ΔI_{qp}. Clearly visible are the Shapiro steps which coincide with the respective FFO-voltages as indicated in the graph. The quasiparticle current rise is read at a bias voltage of in this case 2.5 mV. Data adopted from [Kar13].

Recording many mixer current-voltage characteristics at different FFO bias voltages yields the response of the RF circuit on the frequencies covered by the FFO voltage sweep. Multiple current-voltage curves depicted in one single graph however quickly result in an unclear image. Evaluating a large set of mixer current-voltage characteristics with varied FFO bias conditions yields the spectral behavior of the circuit. Therefore, the quasiparticle current rise ΔI_{qp} of the mixer current-voltage characteristic is evaluated at a given voltage. Since the value of the quasiparticle current increases with increasing radiation, this value contains the information, required to obtain the spectral behavior of the RF circuit. The measured result for this device is depicted in figure 6.12. The measurement points obtained from the measurements presented in figure 6.11 are marked by circles. The values for ΔI_{qp} are evaluated at a voltage of $V = 2.5$ mV which is located on the first photon step. The FFO control line current was set to $I_{\mathrm{CTRL,FFO}} = 11$ mA and the bias current was varied $I_{\mathrm{bias,FFO}} = 8.5$ .. 21 mA. The critical current of the mixer junction without radiation is also illustrated as the black horizontal line. Clearly visible is the quasiparticle current rise to even overcritical values. Since the FFO is operated in the Fiske regime, not all frequencies can be achieved. Therefore, the gaps in the spectrum appear. Nevertheless, a clear frequency dependence of the accepted power can be recognized. Starting at $\nu = 230$ GHz, the power level increases up to a first maximum at 260 GHz. After a minimum at 310 GHz, the next maximum is reached at 335 GHz. Another maximum appears at 410 GHz. This frequency dependence of the transmitted radiation power is attributed to the transmission characteristics of the RF circuit. Therefore, the comparison between this measurement and the simulation yields information about the validity of the simulations and calculations. The simulation result for this circuit is depicted by the red line and clearly follows the measured transmission maxima at 250 GHz and at 410 GHz with

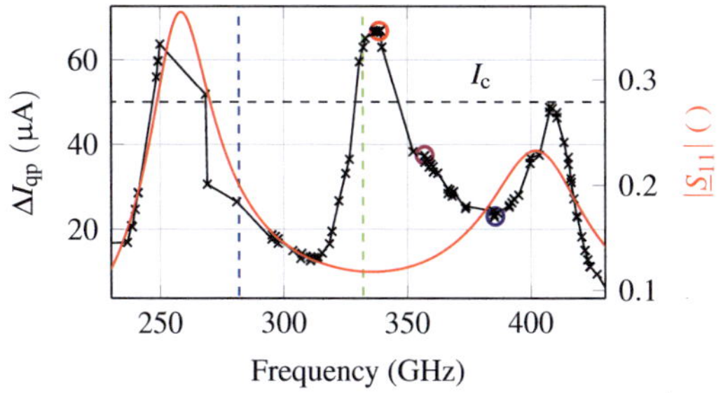

Figure 6.12: Transmission spectrum of one of the realized devices. The graph depicts the quasiparticle current rise ΔI_{qp} depending on the frequency as the black curve. Clearly visible are three distinct maxima at 260 GHz, 340 GHz and 410 GHz. The graph furthermore contains the simulated transmission from the FFO to the mixer depicted in red. An excellent agreement can be seen for the outer maxima. The central maximum does not appear in the simulation. The critical current of the mixer is depicted by the horizontal dashed black line. The calculated values for the resonance frequency of this circuit are depicted by the blue and green dashed lines, calculated with and without the kinetic inductance respectively. Data adopted from [Kar13].

a negligible frequency deviation of less than 2 %. However, the center maximum does not appear in the simulation at all. The calculated resonance frequency for this structure shown above is depicted by the dashed blue line.

If the calculation is redone disregarding the kinetic inductance of the niobium, the result is a resonance frequency of 332 GHz. This is illustrated by the green dashed line and apparently gives the correct frequency for the center maximum which does not appear in the simulation. This indicates the coexistence of two transmission channels, the Cooper pair carried RF-power transmission and the power transmission by quasiparticles.

In total, the mixer is clearly pumped overcritically at multiple frequencies. This overcritical pumping is not required for the actual mixer operation, however, it is most valuable information for the investigation of the coupling between mixer and the FFO. In a real receiver, the most crucial value is the power accepted by the mixer from the receiving antenna. Since the FFO and the antenna have to share a common transmission line towards the mixer, a T-junction is inevitable. Since this T-junction can not be matched to all ports, the high FFO-power allows a weighting of the T-junction, increasing the transmission from the antenna to the mixer, resulting in a higher receiver sensitivity.

Conclusion

A test device, realized in pure niobium technology, is presented. These prototypes comprise a FFO, a DC break in the wiring layer and a mixer with the classical stub tuning approach. The rather large lateral dimensions of all structures in this design are due to the goal of being able to realize the entire chip with photo lithography. The obtained devices could demonstrate a successful transmission of radiation from a flux-flow oscillator to the mixer junction, with an excellent agreement between the simulated and the measured resonance frequencies with even overcritical pumping levels at frequencies around 350 GHz. The high pumping levels are beneficial for the design of fully integrated receivers, which can be used for the realization of weighted T-junction, favoring the incoming signal from the antenna for high sensitivity levels. Due to the excellent agreement between simulation and the measurement, the assumption of valid material parameters and valid simulation techniques is concluded, which in this case concerns the FFO output impedance along with the DC break properties and the mixer tuning. During measurements however, it became obvious that these devices are very sensitive to the direction of how they are connected to the dipstick. Since all ground connections are combined on the ground electrode on the mixer chip, this behavior is most likely due to ground loops resulting from this connection. In order to obtain more flexibility, subsequent designs will also incorporate a DC break in the ground electrode.

6.2.2 NbN-based mixer junction pumped by a flux-flow oscillator at 600 GHz

The conclusions drawn from the previously shown 400 GHz prototype are adopted to the next prototype, which is now based on a NbN/AlN/NbN trilayer on a sapphire substrate. Using this material system, it should be possible to achieve higher frequencies than compared to the pure Nb-based trilayer. However, designing these circuits for higher frequencies highly increases the challenge to the simulation and the fabrication technology. Therefore, this first NbN-based prototype is targeted to operate at frequencies around 600 GHz. The equivalent circuit of the investigated device is depicted in figure 6.13. As compared to the 400 GHz prototype, the entire circuit needs to be redesigned due to the different material system and the different target frequency. Furthermore, a few fundamental changes have been included in this design. These changes are depicted in color in the equivalent circuit. The most obvious difference is the inclusion of the DC break in the ground electrode. Furthermore, a mixer control line was incorporated in order to suppress the Josephson current in the mixer junction. The mixer and tuning circuit is not depicted by its particular components, since different tuning variants have been studied with this prototype. Furthermore, the sizes of the mixer junctions are reduced dramatically, in order to obtain smaller mixer capacities. This results in higher normal state resistances of the junctions, which is why an additional impedance transformation is required in order to match the mixer resistance to the RF network.

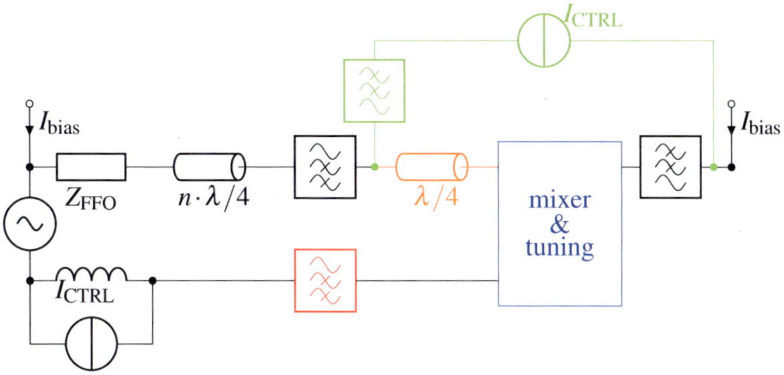

Figure 6.13: Equivalent circuit, depicting the FFO connected to a SIS mixer in niobium nitride technology. The components which are changed in comparison to the previous prototype are depicted in color. The DC break in the wiring layer is implemented as well as a separate $\lambda/4$ transformer right before the mixer. A mixer control line is furthermore included in the design in the wiring layer.

Not only the geometrical restrictions to the mixer are changed, but those of all components and layers. The patterning limits as demonstrated in section 4.4 using electron-beam lithography apply, adding more degrees of freedom to the design. This for instance affects the FFO output impedance, which is now driven to higher values in order to facilitate the impedance matching afterwards. By reducing the FFO idle region width to $W_2 = 2\ \mu m$, the FFO output impedance is increased to $Z_{FFO} = 2\ \Omega$. This value is based on a critical current density of $j_c = 6\ kA/cm^2$ resulting in a capacitance of $C'_{tri} = 58.6\ fF/\mu m^2$. Figure 6.14 depicts an SEM image of the FFO connection to the RF network. In the right hand part of this figure the projection of the two-step isolation layers through the wiring layer can be recognized. This can be recognized by the two steps at the edge of the FFO region. In this prototype design, the impedance transformation was investigated. Therefore, the FFO impedance transformer was realized as a single-stage $\lambda/4$ transformer, as well as a Chebychev type matching transformer. This is indicated in figure 6.13 by the $n \cdot \lambda/4$ transmission line component.

Furthermore, compared to the previous prototype based on pure niobium electrodes, now the material parameters of the niobium nitride and aluminum nitride have to be considered. The transmission line network connecting the FFO to the mixer is selected with a characteristic impedance of $Z_0 = 20\ \Omega$ which is realized by a $11\ \mu m$ wide microstrip line and allows the realization of the subsequent DC breaks.

The wiring layer DC break can be adopted from the previous design, however, it needs to be adjusted to the shorter wavelength. The shorter wavelength is primarily due to the higher frequency and further decreased by the larger kinetic inductance of niobium nitride, which further shortens the tuning relevant parts.

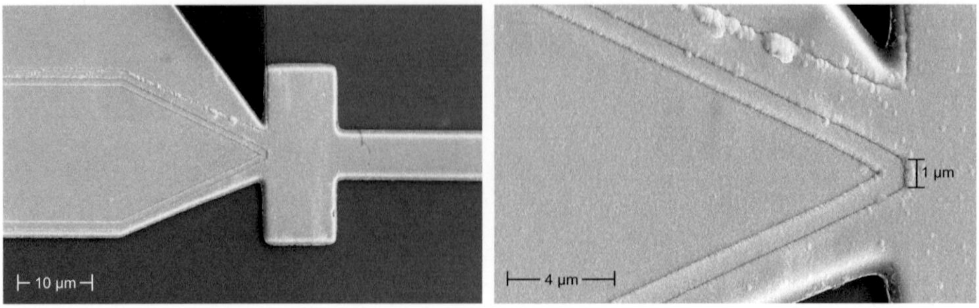

a) b)

Figure 6.14: SEM image of a FFO connected to a $\lambda/4$ matching transformer. In (a), the FFO can be recognized, tapered towards its output, where the impedance transformer is connected, followed by the microstrip transmission line. The close-up in (b) depicts the magnified region of the FFO output, where it is attached to the matching transformer. The designed width of the FFO at its output was 1 µm.

As already mentioned in section 5.2.2, the design of the ground electrode DC break is realized by a gap where parasitic slit modes can propagate. Therefore, the gap is very narrow in the transfer region of the microstrip, and widened abruptly after leaving the microstrip region. The 20 Ω characteristic impedance of the microstrip can however not be matched to the mixer directly, since the mixer areas are decreased to an area of 0.9×0.9 µm^2, resulting in normal state impedances higher than the characteristic impedance of the transmission line network. Therefore, a separate $\lambda/4$ tuner is implemented right before the mixer junction. The higher target frequency of this prototype along with the higher kinetic inductance increases the complexity of the mixer tuning. Therefore, the focus of this prototype is extended towards the tuning method. Whereas in the previous prototype the classical stub tuning has proven to work as expected, this design also includes the other tuning methods presented in section 5.2.5. These different approaches are depicted in the SEM images in figure 6.15. The simplest approach, the classical stub tuning is depicted in the very left, comprising of the mixer, a short piece of transmission line and a radial stub with a wide opening angle. The twinJJ tuning is depicted in the center images. The tuning structure itself only incorporates the two mixer junctions, however, in order to provide a DC connection, the microstrip is extended towards the right end of the image on a length of $\lambda/4$ and is then terminated by a butterfly stub designed to realize a short circuit. This connection hence should not influence on the mixer tuning, since the short circuit of the radial stub is transformed into an open circuit at the mixer by the $\lambda/4$ transformer. The hybrid end-loaded stub tuning is realized in the same manner. It is depicted in the very right of figure 6.15 and includes two butterfly stubs. The left one is responsible for the creation of the tuning capacitor connected through a short

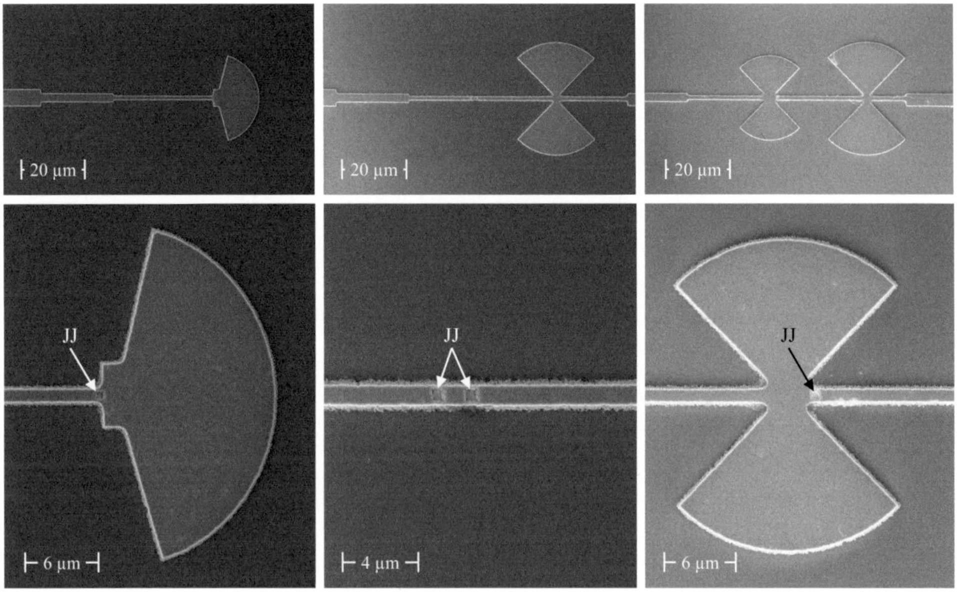

Figure 6.15: SEM images of the different tuning approaches investigated with these prototypes. The left image shows the classical stub tuning, the mixer can be recognized in the close-up at the transition of the narrow microstrip transmission line to the lower impedance, wider microstrip line. The twin junction tuning approach is depicted in the center and the right image shows the hybrid end-loaded stub tuning approach. Note the different scales in the close-ups.

piece of transmission line to the mixer. Also visible is the shorted $\lambda/4$ transmission line for the DC connection. The different tuning methods along with the different matching transformers have been realized multiple times on a $10 \times 10\,\text{mm}^2$ sapphire substrate, including a variation of the tuning length for each permutation.

Figure 6.16 depicts an exemplary SEM image of one of the prototype devices. In the left part of the image, the tail of the FFO is visible, attached to an impedance transformer, in this case a single $\lambda/4$ transformer. This matching transformer is followed by the DC breaks. First, the one in the ground electrode, where the pockets surrounding the microstrip are visible. Following on the DC break of the wiring layer, another matching transformer is visible, transforming the $20\,\Omega$ impedance to a higher impedance of $50\,\Omega$. This $50\,\Omega$ part is the environment, where the mixer tuning is realized. In the SEM image, a hybrid end-loaded stub type of tuning is depicted.

The butterfly stub realizing the tuning capacitor is attached to the $50\,\Omega$ microstrip line, which then leads to the mixer junction. Due to the connection of the mixer towards the right ends of the image, the hybrid end-loaded stub type and the twin junction type tuning can be biased directly through this connection.

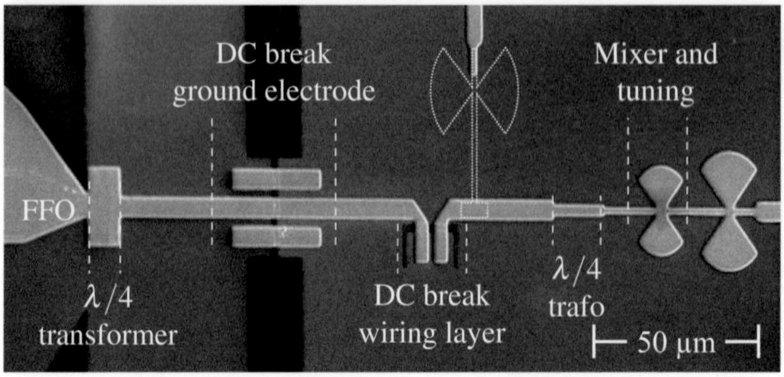

Figure 6.16: SEM image of an entire design realized with a hybrid end-loaded stub tuning. The FFO in the left is followed by the matching transformer, the ground-electrode DC break and the wiring layer DC break. The $\lambda/4$ transformer can be seen right before the mixer region. The dotted white line depicts the site, where an additional DC connector can be attached for the investigation of the control line.

However, the classical stub tuning requires a separate DC connection, which is realized by the connection of a mismatched transmission line terminated by a short circuit i.e. butterfly stub. This can be realized as depicted in figure 6.16 by the white dashed line. In case of the twin junction and hybrid end-loaded stub tuning, this additional connection can be used as a separate connection for the utilization of the wiring layer as the mixer control line. In order to rule out a negative influence of this additional connector, it is not present initially in the twin junction and hybrid end-loaded stub design. It is intended to be added to the structure after the initial investigation of the structure by a lift-off process. In the classical stub tuning with a single radial stub, this connector is realized together with the rest of the wiring layer since it is necessary for the biasing of the junction.

The mixer and the FFO are initially investigated regarding their DC properties. Figure 6.17 depicts the measured current-voltage characteristics of one of the $20 \times 240\ \mu m^2$ flux-flow oscillators measured with variation of the control line current. Clearly visible is the significant increase in the achievable frequency range as compared to the niobium-based FFO. Another observation resulting from these curves is the fact, that the Fiske regime is less pronounced as compared to the niobium-based FFO. The transition to the flux-flow regime is not definitely recognizable. This may be due to the shorter wavelengths that result from the NbN electrodes as well as the higher damping of NbN compared to Nb. However, the flux-flow regime can cover a frequency range up to 1.1 THz as expected. Another feature seen in this set of current-voltage curves is the onset of back-bending of some of the current-voltage curves. This means, the current-voltage curve exhibits an area with a negative differential resistance. Typically, such back-bending of the current-voltage curve is observed in high-

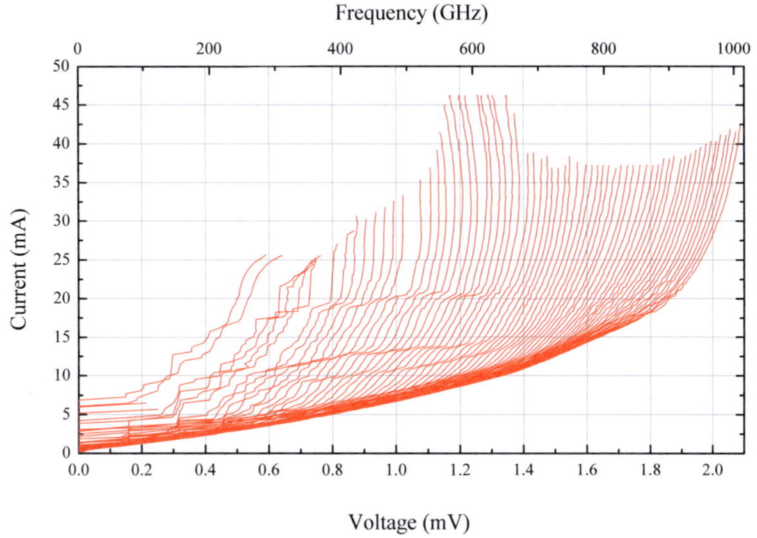

Figure 6.17: Set of current-voltage characteristics of a NbN-based FFO. As expected, the achievable voltages and thus the operation frequency is higher than for Nb-based FFOs. In this case, the highest FFO voltage has been measured at 2.1 mV corresponding to an operation frequency of 1 THz. The Fiske regime can not be distinguished as precisely from the flux-flow regime as for the Nb-based FFO. Data adopted from [Boh17].

j_c short junctions, where the heating of the junction in the voltage state is responsible for this behavior. However, in this case, the back-bending disappears at higher powers, which is why the heating can be ruled out as the origin of this slope. Such curves have already been observed by other groups [105], however, the origin and the implications yet need to be ascertained. Apart from this behavior, the flux-flow oscillator can be operated as expected with a maximum output frequency of more than 1 THz.

Apart from the mixer DC current-voltage curve, the investigation on the mixer now also includes the determination of the proper control-line operation, i.e. the full suppression of the Josephson-current in the mixer. As depicted in figure 6.18a, the control line is attached to the 20 Ω RF network right after the wiring-layer DC break. The dark appearance of the left part of the DC break is due to charging effects during imaging. The microstrip line points towards the mixer on the right. The shallow edges of the structure are clearly visible (see section 4.4), which is due to the lift-off technique. By applying a DC-current through the control line, the Josephson current should be suppressed. The measurement of the critical current dependence on the control-line current is depicted in figure 6.18b. Obviously, the current can be suppressed only to about 75 % of the maximum value of the critical current corresponding to a control-line current of 6 mA.

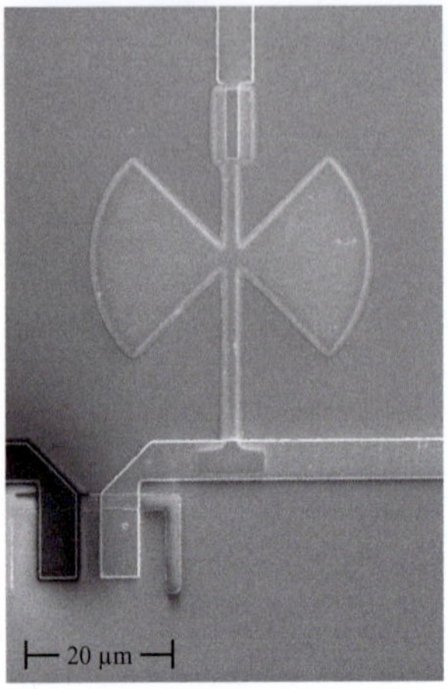

a)

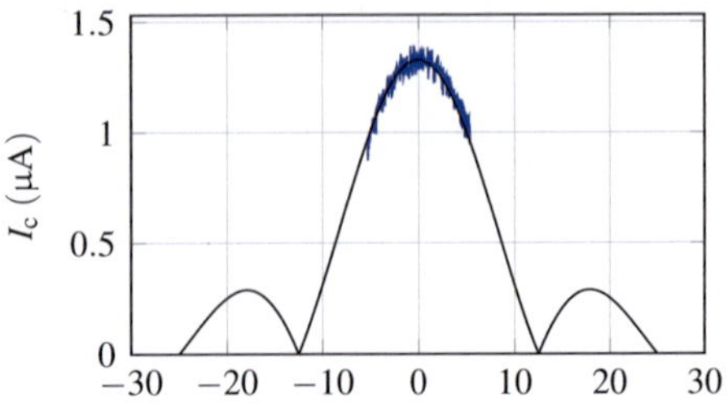

b)

Figure 6.18: (a) A SEM image, showing the additional DC connector for the operation of the mixer control line. Since this connector was added in a separate fabrication step, it is structured using a lift-off technique. The edges are therefore much shallower than those of the transmission lines, which were patterned by etching. (b) The influence of an applied control line current to the critical current of the mixer junction is shown by the blue curve. The critical current can clearly be suppressed to approximately 71 %. However, the control line current can not be increased further, since the critical current density of the control line is exceeded. The black curve depicts a Fraunhofer fit. Data adopted from [Boh17].

This corresponds to the critical current of the control line. Along with the measurement, a Fraunhofer fit is depicted in figure 6.18b for the estimation of the minimum required control-line current for a full suppression of the Josephson current. Apparently, the current needed to be at least twice higher to reach the first minimum of the critical current modulation. In some cases, this is even not sufficient, but an operation point exceeding the fourth minimum might be required [1]. Therefore, a far higher current through the control line is required. However, since the geometry of the control line is dictated by the RF network, the width can not simply be increased, hence, this approach does not work as expected.

The DC characterization of the mixer itself results in a mixer current-voltage characteristic as depicted in figure 6.19 by the blue line for (a) a single mixer junction and (b) a twin junction device. The normal state resistances extracted from these curves exhibit values of 180 Ω for the single junctions and 90 Ω for the twin junction tuning respectively. This resistances are significantly higher than what was expected from preliminary investigations on this tri-layer. The pumping levels of these devices are thus expected to be low due to the resistance mismatch of the mixer to the RF environment. Figure 6.19 also contains exemplary mixer current-voltage characteristics with applied FFO power of $\nu = 603$ GHz and $\nu = 542$ GHz

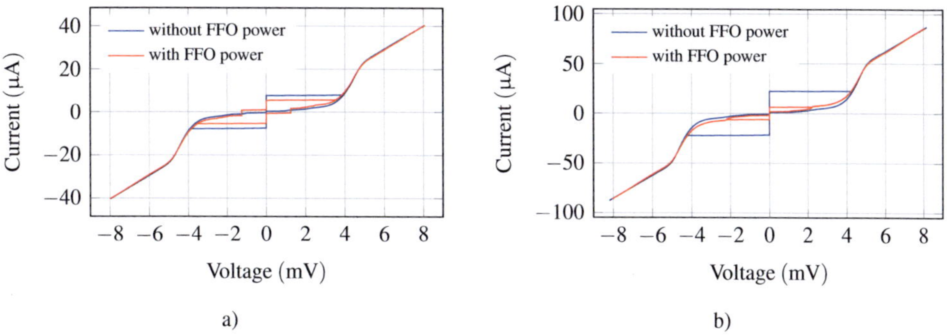

Figure 6.19: Exemplary current voltage curves of (a) a hybrid end-loaded stub tuning design and (b) a twin junction tuning design with and without FFO power. The normal state resistances extracted from these curves are 180 Ω for the single junction designs and 90 Ω for the twin junction design. Data adopted from [Boh17].

respectively. As expected, the pumping levels are not as high as e.g. in the niobium-based prototype. However, the frequency behavior is expected not to be shifted due to the resistance mismatch. This is verified by simulations, which apart from a larger bandwidth show no deviations compared to the low-resistance simulations.

In order to investigate on the frequency properties, again, the FFO-induced quasiparticle current ΔI_{qp} in the mixer is monitored for different FFO bias points. At first, the question of the correct representation of the FFO in the simulation is addressed. This question condenses to the selection of the correct port in the simulation. As discussed in section 5.2.1, the available options are a wide port, covering the entire width of the subsequent matching transformer or a narrow port, following the lateral extension of the FFO at its output. Both approaches have been realized on the prototype chip and exemplary measurements are depicted in figure 6.20a and 6.20b. These figures show measurements of the quasiparticle current rise versus the frequency as the black curves. In both cases, a Chebychev matching transformer was realized after the FFO and the identical hybrid end-loaded stub tuning circuit was used and optimized for 600 GHz employing a tuning length of $\ell_{\mathrm{MS,tune}} = 2.45$ µm. These designs hence only differ in the port width, that was used to create the matching transformer. The design, created with the wide port (6.20a) significantly deviates from the simulation, which is shown by the red curve, whereas the design employing the FFO-like port width (6.20b) closely matches the simulation, with barely any difference in the peak transmission frequency of in this case 600 GHz. By the correct port selection for the FFO, the operation frequency can thus be simulated quite closely for the hybrid end-loaded stub tuning. However, the simulated bandwidth does not yet match the measured one, the origin of which yet has to be identified and investigated in the future.

The result on the identical design with a single stage matching transformer instead of the Chebychev-type transformer is depicted in figure 6.20c. Apart from a slight deviation in peak transmission frequency, the single stage transformer enables nearly the identical frequency behavior as the Chebychev matching transformer. The peak transmission frequency deviates by only ≈ 3 % and the maximum FFO induced quasiparticle current $\Delta I_{\mathrm{qp,max}}$ is in both cases at 1.4 μA. This concludes that both matching transformers can be utilized to realize the required impedance transformation to match the flux-flow oscillator to the higher impedance transmission line network. The wider bandwidth of the Chebychev type transformer can be beneficial, however, it is hardly measurable, since the bandwidth limiting element is in theory the tuning of the mixer capacitance. Therefore, the wide bandwidth of the Chebychev transformer only gains significance, if the frequency of the mixer tuning deviates from the designed frequency. The representation of the FFO in the simulation is thus established, with two options to match the low impedance FFO to the RF network. The mixer tuning that was used in the previous investigation was the hybrid end-loaded stub tuning exclusively, in order to allow a fair comparison of only the FFO and the matching, without the tuning influencing on the result.

In addition to devices probing the matching of the FFO, the chip also contains designs with different mixer tuning approaches, enabling the comparison between the tuning approaches. These designs are furthermore realized on chip multiple times, each with a different tuning length, in order to obtain data from more than just a single prototype. All devices were designed to operate around 600 GHz with a tuning length induced variation of approximately 20 GHz in the case of the hybrid end-loaded stub type and 50 GHz in the case of the twin junction tuning design. In the case of the classical stub tuning, the frequency variation due to the different tuning lengths is designed to be 30 GHz. The single stage matching transformer is used in all the results showed in the following graphs. Figure 6.21 depicts exemplary measurement results of the three different tuning approaches.

Figure 6.21a contains the result for a classical stub tuning with a tuning length of only 620 nm. The design frequency of 600 GHz is not achieved, but the highest RF transmission is measured at approximately 550 GHz. The origin of this deviation is most likely attributed to this very short tuning length which results in increased influence of fabrication inaccuracies and parasitic reactances on the resonance frequency. This is attributed to the large kinetic inductance of the niobium nitride, resulting in shorter wavelengths and thus shorter tuning lengths. As can be seen in figure 6.15, the width of the tuning microstrip line is already at larger values. This is the attempt to increase the tuning length (see equation 2.52) due to the lower characteristic impedance. This was optimized by simulations, but due to the discontinuity in width at the transition to the 50 Ω mixer connection, wider transmission

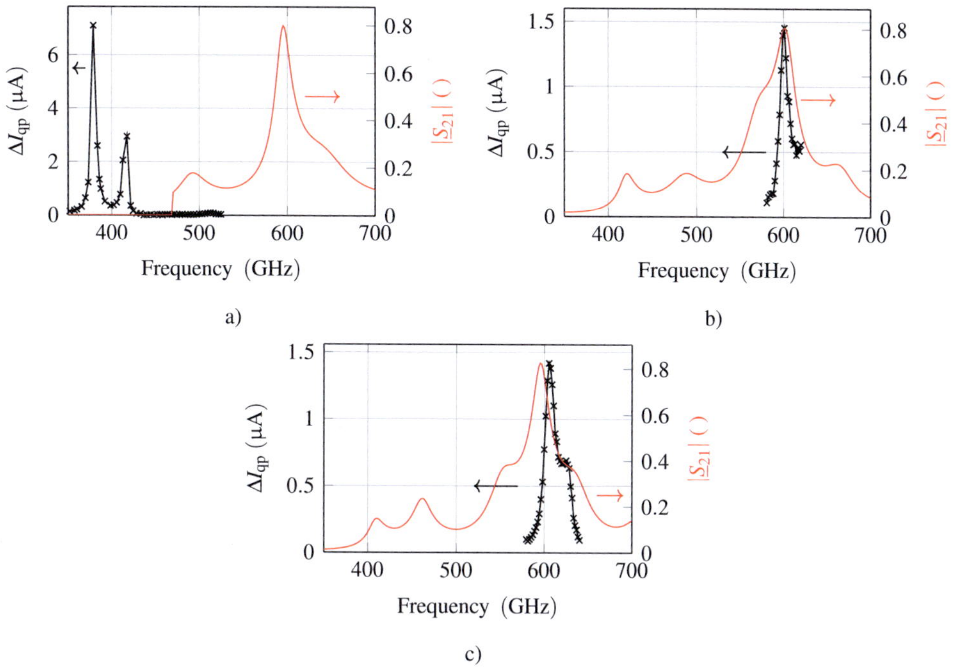

Figure 6.20: Comparison of the simulated and the measured transmission spectra from the FFO to the mixer. The simulation results are shown in red and the measured points are depicted in black. (a) The obtained result for the two-stage Chebychev type matching transformer, simulated with a wide port. (b) The results for the two-stage Chebychev and (c) quarter-wavelength matching transformer, each simulated with a port width that matches the width of the FFO output. Data adopted from [Boh17].

lines do not yield the desired improvement in tuning length. Although, this tuning approach achieves a larger RF power transmission to the mixer than compared to the previous results with the hybrid end-loaded stub tuning, an upper limit in operation frequency seems to be approached due to the short tuning lengths. Therefore, the other approaches discussed in section 5.2.5 are beneficial, since they allow for longer tuning lengths due to their intrinsic setup.

A result of the hybrid end-loaded stub has already been depicted in figure 6.20c, with a tuning length of 2.45 µm. The tuning with a length of 2.23 µm is shown in 6.21b. Similar to the measurement result presented previously, a slight frequency deviation of the design to the measured peak frequency is visible, but very small of about 17 GHz. The FFO induced quasiparticle current at this peak frequency is again at 1.5 µA, which is identical to the previously shown measurements. An exemplary measurement result for the twin junction tuning approach is depicted in figure 6.21c.

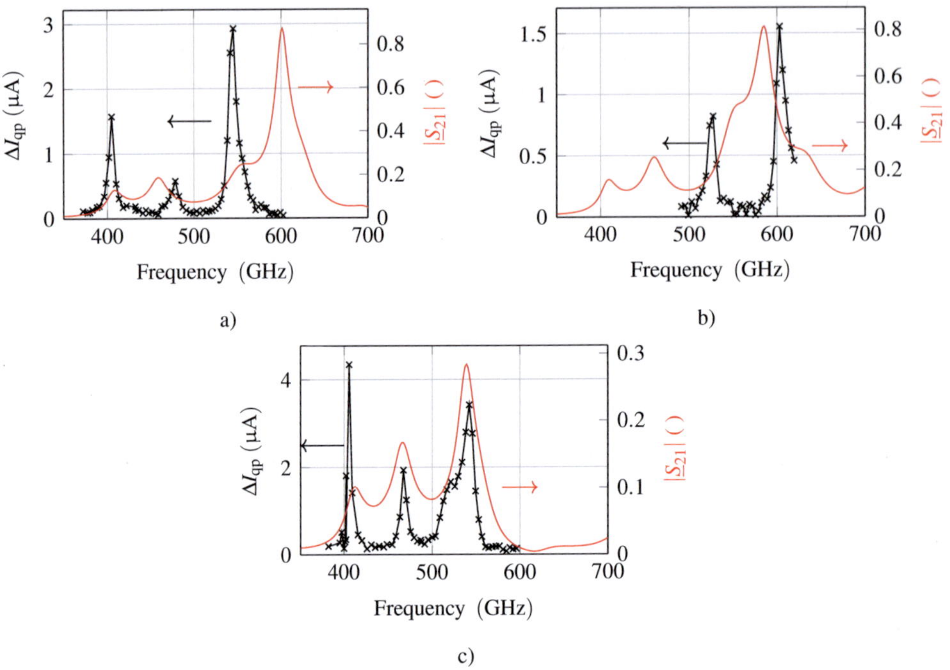

Figure 6.21: Comparison of the simulated and the measured spectra of power transmission from the FFO to the mixers with different tuning approaches: (a) The case of the classical stub tuning; (b) The result of the hybrid end-loaded stub tuning; (c) The results of the twin junction approach. Data adopted from [Boh17].

A clear conformity of the peak frequency between the measured values and the simulated ones can be observed. This close match between the simulation and the measurement indicates a good agreement between the assumed material parameters and the realized ones. This also leads to the assumption, that fabrication inaccuracies are responsible for the slight deviation in peak transmission frequency of the hybrid end-loaded stub tuning approach. Especially at the transition of the butterfly stub to the microstrip line, the proximity effect of the lithography induces a rounding of the corners which in the simulations are assumed sharp. This might be the reason for the small deviation in peak transmission frequency. Nevertheless, both, the twin junction as well as the hybrid end-loaded stub tuning show good agreement between the simulation and the measurement at operation frequencies of higher than 600 GHz, at least for these single exemplary devices shown above.

In order to compare multiple devices with their respective simulations, the peak transmission frequencies are considered. This consideration does not only include the frequency with the maximum power transmission, but also the possible side maxima in the spectrum. The spec-

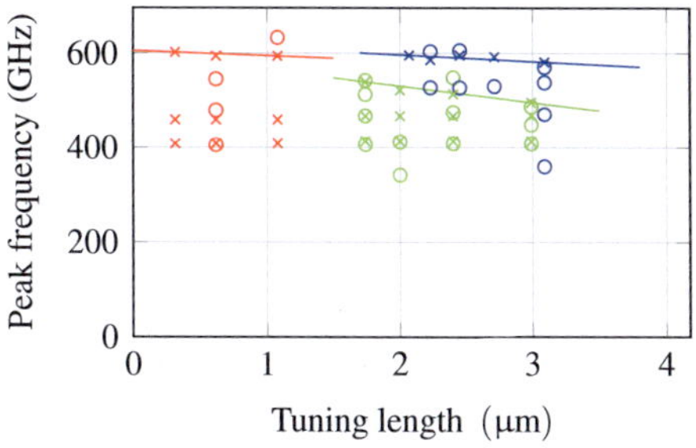

Figure 6.22: Comparison of the peak transmission frequencies of the different tuning approaches for multiple devices. The measured peak transmission frequencies are depicted as open circles, the simulated ones are given as crosses. The classical stub tuning is depicted in red, the twin junction tuning is shown in green and the hybrid end-loaded stub tuning is colored in blue. Data adopted from [Boh17].

trum in 6.21c for instance contains three maxima, which are considered. The frequencies of these maxima are noted and depicted over the tuning length in figure 6.22. The measured transmission maxima are hereby represented by the open circles, the maxima derived from the simulations are depicted as crosses. Clearly visible are the shorter tuning lengths for the classical stub tuning compared to the other tuning approaches at the same frequencies. For each tuning length where crosses appear, a device has been realized. Whereas the classical stub tuning barely yields any match to the simulation, especially the twin junction tuning yields a good agreement for almost all devices even including the side maxima. The expected slope for all tuning approaches can be recognized in this figure as well. A longer tuning length results in a lower resonance frequency. However, some of the maxima seem independent of the tuning length. These features hence are expected to be due to parasitic resonances in the RF network. In total, the tuning approaches behave as expected, with an excellent frequency match between simulation and measurement for the twin junction tuning. The short tuning lengths of the classical stub tuning is approaching a limit in operation frequency due to the high kinetic inductance of the niobium nitride electrodes. Therefore, the alternative approaches prove to be extremely beneficial, boosting the tuning length for the same frequency to larger values and thus enabling higher operation frequencies.

Conclusion

The development of a NbN-based 600 GHz prototype device is presented comprising a flux-flow oscillator connected to a SIS mixer through the required impedance transformations and DC breaks in both metalization layers. The conducted DC measurements showed the FFO operation up to approximately 1 THz, which is possible due to the usage of niobium nitride as the electrode material. Using these prototypes, the correct FFO representation in the simulation was successfully investigated. All three mixer tuning approaches presented in chapter 5 were investigated, resulting in a perfect match between the simulated peak transmission frequencies in the case of the twin junction tuning. The hybrid end-loaded stub also yields a close match between simulation and measurement. The classical stub tuning approach also proved to be operational, however, the deviation between simulated and measured spectra was increased significantly compared to the other approaches. Furthermore, this approach exhibits strong limitations in frequency, due to the high kinetic inductance of the NbN. Therefore, the hybrid end-loaded stub tuning and the twin junction tuning approach are advantageous. This prototype furthermore contained a mixer control line in the wiring layer in order to enable the suppression of the Josephson critical current in the mixer. However, the current required for a full suppression of the Josephson critical current exceeds the critical current of the control line in the wiring layer, hence a suppression of the critical current of the SIS junction was not possible. Due to an unexpected high mixer normal state resistance, the impedance matching between the mixer and its RF periphery is not accomplished, which is the reason for overall low pumping levels. This can however be improved in future devices by the usage of an additional impedance transformation. Except for the mixer control line, this preliminary device proved all components to work as expected, hence, including the antenna investigation in the first section of this chapter, all components that are required for the realization of an integrated receiver have been successfully tested. Therefore, the complexity of the RF circuit of the devices in the following chapter is increased further, resulting in a fully integrated receiver.

7 Fully-integrated superconducting receiver

The gathered results of the prototype devices are used to realize a fully integrated receiver. This device includes the FFO integrated with the SIS mixer as well as an antenna to receive an external signal. Due to the findings yielded by the prototype devices, the design has to be altered slightly. The target frequency for this integrated receiver was further increased to 750 GHz. This frequency selection was done without the planning of a particular application. In fact, the frequency range surrounding 750 GHz is of no practical interest in ground based applications, since the atmosphere's transmission is absorbing nearly all radiation in this frequency range. Nevertheless, this frequency would prohibit pure niobium based integrated receivers with FFOs to work, since the operation frequency exceeds the gap frequency of niobium. On the other hand, the operation frequency was not selected to higher levels in order keep the challenges to the design and fabrication comparable to the test devices.

7.1 Design changes as compared to prototypes

Due to the findings from the preliminary devices, the design has to be altered, to yield the desired properties. A trilayer of 8 $\mathrm{kA/cm^2}$ is used, which enables the further reduction of the junction size, without increasing the normal state resistance. Two different junction sizes were used in these integrated receivers: $0.7 \times 0.7\ \mu\mathrm{m}^2$ and $0.9 \times 0.9\ \mu\mathrm{m}^2$. This leads to normal state resistances of $R_\mathrm{N} = 150\ \Omega$ and $85\ \Omega$ respectively. The RF circuit is again designed to exhibit a characteristic impedance of $Z_0 = 20\ \Omega$. Overcoming the low power transmission to the mixer seen in the 600 GHz prototype, is achieved by an additional impedance transformation. Figure 7.1 depicts the equivalent circuit of the considered structures. Note, that this additional impedance transformer is not depicted separately, but is included in the tuning block.

A significant change in the design results from the inclusion of an antenna into the circuit, which requires the implementation of a T-junction. A twin-slot antenna is used which is designed to exhibit an output impedance of $40\ \Omega$.

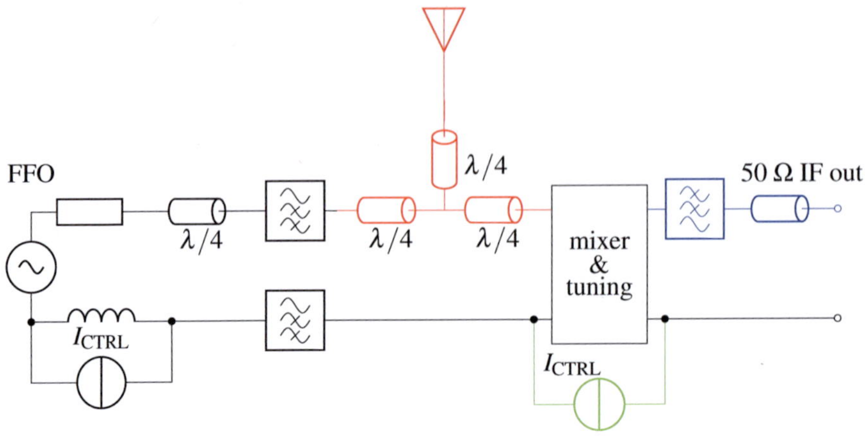

Figure 7.1: Equivalent circuit of the fully integrated receiver. The components that have been changed compared to the previous 600 GHz prototype are highlighted in color.

In order to achieve this impedance, the antenna incorporates a $\lambda/4$-matching transformer. The 40 Ω microstrip line is then connected to one of the inputs of the T-junction. The other input is fed by the flux-flow oscillator. The matching transformer at the flux-flow oscillator is a single-stage $\lambda/4$ transformer, yielding a characteristic output impedance of 20 Ω. The single-stage transformer was preferred over the Chebychev transformer in this case, since these frequencies require the first stage of the Chebychev transformer to be extremely short. Therefore, the simulation results using the single-stage transformer exhibited a higher power transmission to the 20 Ω microstrip line. The T-junction needs to combine these signals onto the single 50 Ω output, which is directed to the mixer. From the 50 Ω transmission line, the impedance is transformed using the additional $\lambda/4$ transformer, to obtain a close match to the mixer normal state resistance.

Since the approach of realizing the mixer control line in the wiring layer did not yield the required suppression of the Josephson critical current, the control line is now realized in the ground electrode as discussed already in chapter 5. This implies further severe design alterations that can be seen in the micrograph of the fabricated structures depicted in figure 7.2. The bright areas represent the niobium nitride of the wiring layer, the green areas are the ground electrode. The green shade to the niobium nitride of the ground electrode is added from the silicon monoxide on top of the ground electrode. The image depicts the center part of a design with the FFO tail visible in the left part of the image, followed by both DC breaks before entering the T-junction in the center of the image. The twin-slot antenna can be seen left from the T-junction as the two blue slots in the ground electrode. The transition from the slots to the microstrip transmission line is achieved by a transverse microstrip line across the slit, terminated by a short-circuit. This short circuit is again realized by a radial stub. The microstrip lines from both slots join in the center of the antenna, after which a $\lambda/4$ transformation achieves the 40 Ω output impedance. The antenna signal is then fed to the second input of the T-junction. After the signal combination, both signals are fed to the mixer junction embedded in the hybrid end-loaded stub tuning structure.

For a control line realized in the ground electrode, some sort of gap is required to guide the control line current. In order to realize this gap in the vicinity of the mixer without deteriorating the properties of the microstrip line circuit, the hybrid end-loaded stub tuning approach has to be used. Furthermore, the biasing connection needs to be altered due to the resulting gap in the ground electrode. In previous designs, this connection was realized using a $\ell = \lambda/4$ microstrip transmission line terminated with a butterfly stub. The transmission line transformed the very low impedance of the butterfly stub to an open circuit at the point where the RF-circuit is connected, hence not influencing the properties of the RF network. Due to the gap in the ground electrode, the microstrip connection can not be used in this case, but has to be replaced by a coplanar transmission line setup as shown in figure 5.17. Therefore, the butterfly stub is replaced by a consecutive multiple $\lambda/4$ transformation terminated by a $Z_0 =$ 50 Ω transmission line which in literature is referred to a RF-choke. Since the intermediate frequency is far lower than the design frequency, this connection is used not only for the biasing, but also for the intermediate frequency output. After the choke, the 50 Ω coplanar waveguide transfers the signal via a taper to the IF-periphery. The onset of the taper is visible in the right end of the image.

A close-up of the mixer site can be seen in the SEM image in figure 7.2. In this top view on the wiring layer, the mixer can be recognized as the square shaped indention in the vicinity of the intersection of the butterfly stub and the microstrip. Apparently, the rounding at this intersection is more prominent than compared to the structure shown in figure 6.15. This is not only due to the smaller lateral structure sizes, but might be owed to dose fluctuations during lithography. Connected to the right end of the center of the butterfly stub is the coplanar waveguide coming from the RF-choke. The edge of the gap in the ground electrode can be recognized as the dark conical area with edges parallel to the radial stub edges. The gap of the coplanar waveguide needs to be tapered in order not to deteriorate the properties of the butterfly stub. In an overlay, this figure also contains the design file as dashed lines. The edges of the ground electrode are marked in red, the junction area is limited by the blue line and the wiring layer edges are depicted in green. Clearly visible is the perfect match, of both the dimensions of each layer with the design, as well as the alignment of the layers to each other. Except for the rounded intersections between the stubs and the microstrip line, the fabrication process thus yields the expected design accuracy.

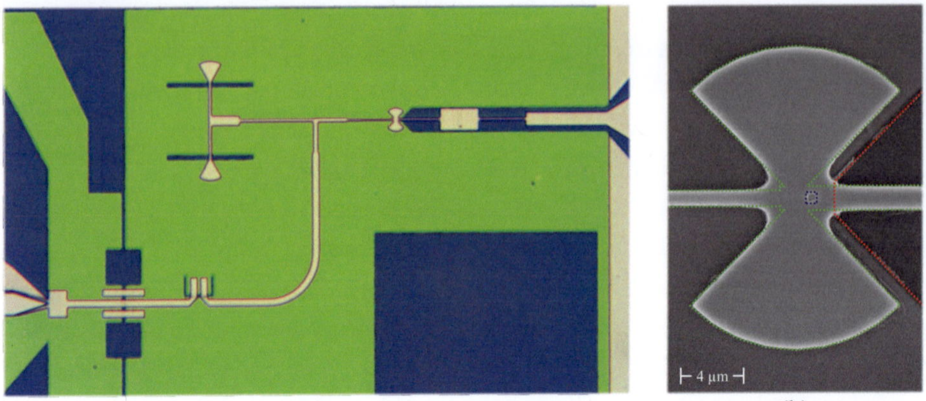

(a) (b)

Figure 7.2: Micrograph and SEM image of the integrated receiver design. The micrograph of an integrated receiver is depicted in (a). The output of the FFO can be seen in the lower left corner attached to a quarter wave matching transformer, followed by both DC breaks. The double slot antenna is located in the top left part of the image. Both signals are combined in the T-junction, located in the center of the image and then fed to the mixer. The IF output of the mixer is realized by an RF choke in quasi-coplanar waveguide technology, which can be seen in the top right part of the image. (b) SEM close-up of the mixer and tuning attached to the intermediate frequency output. In the overlay, the design file is included as the colored dashed lines.

As in previously demonstrated prototypes, multiple designs were realized on a $10 \times 10\,\text{mm}^2$ sized sapphire substrate. Due to the incorporation of the intermediate frequency output taper, the chip area consumed by a single integrated receiver is much larger, than without the intermediate frequency connection. Figure 7.3 depicts the design file of an integrated receiver. This $1.8 \times 1.8\,\text{mm}^2$ sized die contains all necessary connections for the operation of the integrated receiver, including the coplanar intermediate frequency output, located in the top part of the image, as well as the DC-connector pads. The particular purpose of each connector part is given in the figure. With the given dimensions, a single $10 \times 10\,\text{mm}^2$ substrate can hence accommodate 16 dies.

Conclusion

Since the prototype devices yielded excellent agreement between the simulated transmission frequencies and the measured ones for frequencies as high as 600 GHz, the complexity of the RF circuit can be further increased to a fully integrated receiver. The required design changes from the last preliminary device to this integrated receiver are described in this section. The most severe alteration is the inclusion of an antenna, which requires the design and incorporation of a T-junction. The mixer junctions are designed as $0.7 \times 0.7\,\mu\text{m}^2$ and

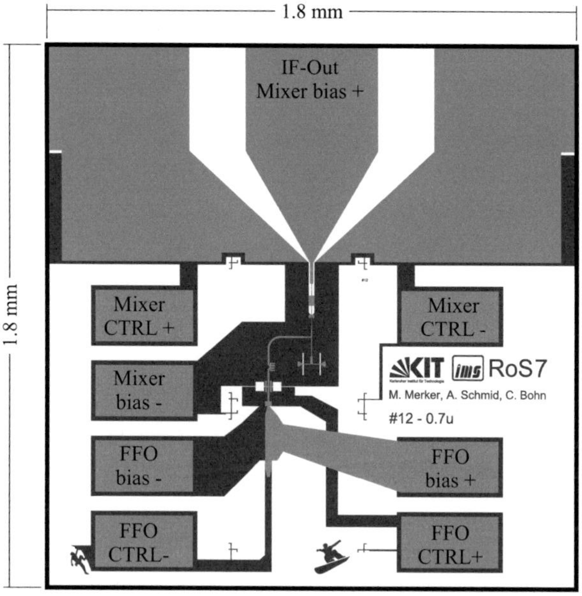

Figure 7.3: Design file of an entire die of the integrated receiver. The wiring layer is depicted in light gray, the ground electrode appears is colored in darker gray. The IF output is located in the top part of the image. The connector pads for the particular DC connections are named in the image. The entire die size is $1.8 \times 1.8\ \text{mm}^2$.

$0.9 \times 0.9\ \mu\text{m}^2$ sized junctions, on an $8\ \text{kA/cm}^2$ trilayer. The high normal state resistance of the mixer is accounted for by an additional quarter wavelength transformer, granting the impedance matching for this high mixer impedances to the microstrip transmission line network.

Furthermore, the mixer control line is realized in the ground layer, which is why the hybrid end-loaded stub tuning approach was used. The intermediate frequency output therefore has to be realized in a coplanar waveguide technology which can directly be connected to the coplanar waveguide on the connecting PCB. Due to this connection, one design requires more space on the chip as compared to the prototype devices without the IF connection. Therefore, a $10 \times 10\ \text{mm}^2$ can accommodate 16 different designs. As done with the preliminary devices, these devices are also initially characterized in a dipstick measurement, investigating on the transmission from the FFO to the mixer.

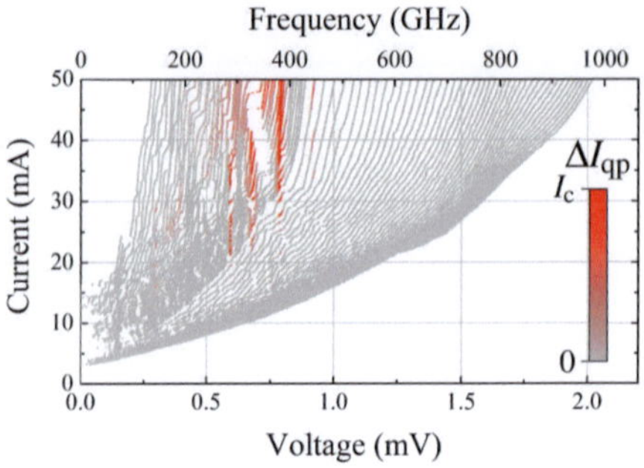

Figure 7.4: Selection of FFO current-voltage characteristics for different control line currents ranging from 7.6 to 41 mA. The induced quasiparticle current rise in the mixer is illustrated in color code. The gray lines represent the FFO operation points which do not result in a measurable quasiparticle current rise. The red areas represent the operating points with a critical mixer pumping.

7.2 Electrical characterization

Prior to the quasioptical measurements in the cryostat, the dies are characterized for their properties in a dipstick measurement. As with the preliminary devices, the spectrum of the transmission properties from the FFO to the mixer can already be recorded without the need for a time-consuming cryostat measurement. Especially, the altered components compared to the preliminary devices are of interest. In particular, the T-junction incorporation also involves a reduction in power transmission from the FFO to the mixer by approximately 3 dB compared to the devices without the T-junction. However, the overall transmission from FFO to mixer is expected to be increased due to the additional $\lambda/4$ transformer, accounting for the mixer impedance matching to the RF network. Furthermore, the redesigned mixer control line in the ground electrode can be investigated with respect to a full suppression of the Josephson critical current.

As in the investigation of the prototype devices, shown in the previous chapter, a current-voltage curve set of the FFO is recorded first, one of which is depicted as an example in figure 7.4 by the gray lines. The color code indicates ΔI_{qp} in the mixer, induced by the FFO operation. For clarity reasons, not all recorded curves are displayed, which is why the maximum frequency is below 1 THz. Following on the FFO characterization, the mixer DC characteristic is recorded without applied FFO power and subsequently with incident FFO power. Since the mixer control line is not yet used for this measurement, the mixer current-voltage characteristic contains both Shapiro and photon-assisted tunneling steps. The FFO bias voltage was set to 0.79 mV, resulting in an emission frequency of 382 GHz. The normal state resistance $R_N = 140\,\Omega$ of this mixer is approximately at the expected value of $150\,\Omega$.

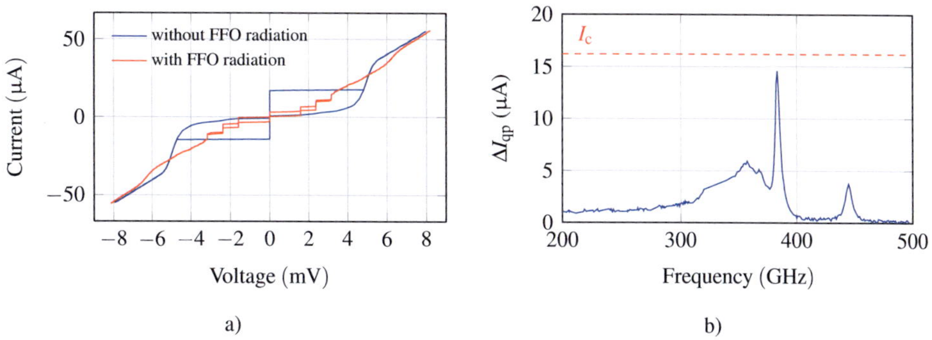

Figure 7.5: (a) A mixer current-voltage characteristic with and without FFO radiation. (b) Transmission spectrum of the FFO to the SIS junction evaluated from the quasiparticle current rise. The critical current level of this mixer is depicted by the dashed red line.

Due to the additional resistance matching, the mixer can accept more power from the FFO as compared to the 600 GHz preliminary devices. This is confirmed by the height of the photon-assisted tunneling step in figure 7.5a. The current right below the gap increases to levels close to the critical current of the mixer. In particular, the mixer is susceptible to the three frequencies 280, 315 and 382 GHz (compare the three red vertical areas in figure 7.4), which was not expected from the simulation since the design frequency was 750 GHz. So the yielded operation frequency is significantly lower than the simulated one. This is also reflected in the graph of the frequency dependence of the induced quasiparticle current as depicted in figure 7.5b. Clearly visible is the peak pumping level of the mixer at a frequency of 384 GHz. The induced quasiparticle current of 15 µA almost equals the Josephson critical current of the junction. The side maximum towards lower frequencies is at lower pumping levels, as expected from the color code in the FFO current-voltage curve set. This overall behavior is yet to be explained; simulations have not yet yielded a satisfactory explanation of the measurements.

As it will become apparent in the following sections, this was not only a singular device failing to achieve the target frequency, but all of the realized structures did not operate at the 750 GHz design frequency, but at frequencies between 300 and 400 GHz. In order to examine the origin of this deviation, the realized structures were investigated. At first, the structures were inspected for deviations from designed dimensions. However, as depicted in figure 7.2, the realized structures accurately fit the designed dimensions. The only deviation is given by the rounded edges of the butterfly stub at the connection to the microstrip transmission line.

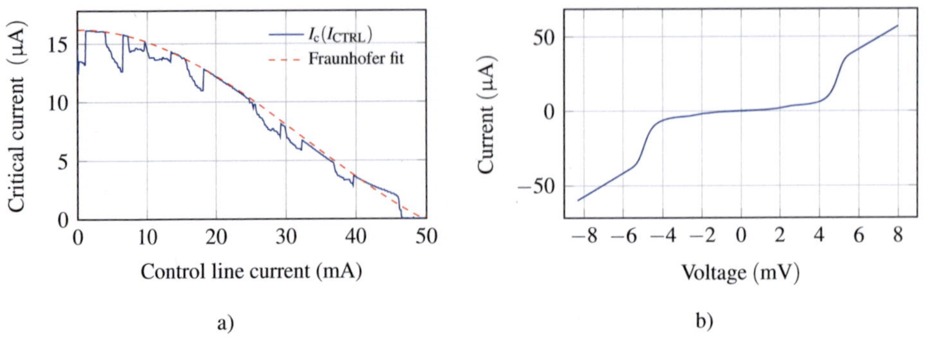

Figure 7.6: (a) The measured modulation of the critical current with the control line current. The Fraunhofer fit is fitted to the assumed envelope of the measured curve. For control line currents exceeding 45 mA, the critical current of the mixer junction is fully suppressed. The resulting current-voltage characteristic of the mixer is depicted in (b).

Therefore, this structure was adopted to CST and simulated with these rounded edges, however resulting in an unaltered simulation spectrum. These rounded edges can thus be ruled out as the origin of the observed massive frequency shift.

Since this behavior could not be reproduced by the simulations, test devices have been fabricated, focusing on various points in the entire RF circuit. Since one of the major design alteration as compared to the preliminary devices from the preceding chapter was the inclusion of the T-junction and the antenna, these components were left out on one test device. This test design however did not show any different behavior other than a lower power transmission. Furthermore, the FFO output impedance was altered. The calculated value hereby served as the central value, around which arbitrary values were selected with higher and lower values than the calculated one. Again, there was no change in the observed microwave behavior of the circuit. Further tests have not yet been conducted, but are necessary, in order to obtain clearance on the origin of the unexpected frequency shift. Nevertheless the device shown above has exhibited an almost critical pumping of the mixer by the flux-flow oscillator.

Apart from the RF properties of the connection between mixer and FFO, the operation of the mixer control line is examined. Figure 7.6 depicts the results obtained from this investigation. The modulation of the Josephson critical current with the control line current is depicted in the left graph as the blue line along with a Fraunhofer fit. Clearly visible is the decay of the critical current with increasing control line current down to zero critical current for control line currents exceeding approximately 47 mA. Comparing this results with those presented in section 6.2.2 results in a lower current efficiency in magnetic field generation by the control line in the ground electrode. This behavior is not unexpected, since the control line in the wiring layer exhibits a significantly larger inductance, due to its narrow linewidth. Additionally, discontinuities are present in this graph. These can result from the shape of the current path (compare figure 5.18). The current concentrated at the edges of the electrode has to pass the edges with small radii at the mixer site. These edges result from the tapering of the coplanar gap so as not to deteriorate the butterfly stub performance by removing its ground electrode. Due to these small radii, current crowding effects have to be considered [147–149], [VIMS15]. This can result in Abrikosov vortices entering the lower electrode, which disturb the Josephson critical current due to their intrinsic magnetic field. A preliminary confirmation of this hypothesis seems to be the observation of the critical current modulation exhibiting some kind of envelope, to which the current jumps back more or less periodically. This envelope is very close to a Fraunhofer fit as indicated by the red dashed line. Of course, no ideal Fraunhofer spectrum can be expected to be obtained by this measurement, since the magnetic field created at the junction is not homogeneous due to the highly non-uniform current along the edge of the bottom electrode gap. The penetration of vortices into the ground layer may be reduced by magnetic shielding and additional filtering, lowering the noise levels. However, if the Josephson critical current is fully suppressed, the vortices should not influence on the mixer performance, since the comparably weak magnetic fields do not interact with the quasiparticles. The complete suppression of the Josephson current by this magnetic field results in a current-voltage characteristic of this device as depicted in figure 7.6b.

Conclusion

This section presents the results obtained by dipstick characterization of the fully integrated receiver. This characterization includes the investigation of the FFO radiation transmission to the mixer junction as well as the characterization of the alternative approach for the mixer control line in the ground electrode. The recorded current-voltage characteristics of the mixer under FFO radiation demonstrated significantly higher pumping levels as compared to the 600 GHz device. However, the power transmission is limited to frequencies significantly below the target frequency. The measured devices all exhibited a measurable transmission at frequencies in between 300 and 400 GHz. This behavior can not yet be modeled by the simulation. Apart from the transmission characterization, the altered mixer control line is investigated, which is now located in the ground electrode. This measurement yielded a full suppression of the Josephson critical current in the mixer junction for control line currents higher than 45 mA, hence demonstrating the successful implementation of the mixer control line. This dipstick characterization did not include any external radiation.

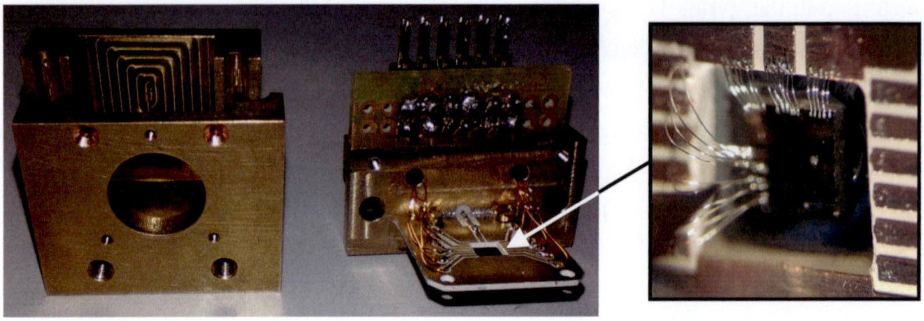

(a) (b)

Figure 7.7: Photographs of the mixer block and the connector board. (a) In the front, the mixer block has an opening for the silicon lens. The connector board is attached to the lid of the mixer block. This lid with the connector board is inserted in the mixer block from the top and the device under test can be bonded to the connector board. A close-up of the mixer glued to the silicon lens and bonded to the connector board is shown in (b). Image adopted from [Rei17].

7.3 Quasioptical measurements

Since a high RF power transmission was demonstrated by the measured devices, a selection thereof was characterized regarding their quasioptical response. For this characterization, a single die is glued on the back of a hyperhemispherical silicon lens which is installed inside a detector block that can be mounted on the cold plate of a cryostat. A Rogers connector board is designed, providing all necessary DC connections as well as the 50 Ω coplanar waveguide intermediate frequency output. Figure 7.7 shows a photograph of the connector board (center part of the image) which is then mounted in the detector block (left part of the image). The connection between the adapter board and the chip is realized by aluminum bonds as depicted in the close-up in figure 7.7. The DC connections are located to the right and to the left of the chip, whereas the intermediate frequency output is located in the top part of the image. This photo was taken during the bonding process.

The setup used in this case is depicted schematically in figure 7.8. A backward wave oscillator (BWO) is used as a signal source with an instantaneous frequency coverage ranging from approximately 180 to 390 GHz. The output power varies between 190 μW at the lower frequency boundary and increases up to values exceeding 9 mW for frequencies around 380 GHz. The achievable frequency range can be increased by using a passive frequency tripler, which is installed at the BWO output. The measured frequency spectrum with this tripler thus ranges from 550 up to 1150 GHz. However, the complex output power spectrum significantly drops to values ranging from 1.5 μW at the lowest frequencies to 5.4 μW at the

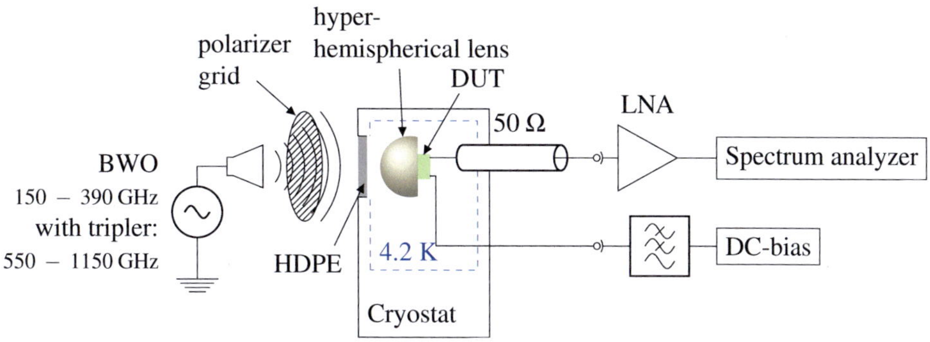

Figure 7.8: Optical setup to determine the optical response of the integrated receiver. The chip inside the detector block is mounted on the cold plate of a cryostat. The mixer is irradiated through a HDPE window in the cryostat by a BWO radiation source. A rotatable polarizer grid is used to enable the attenuation of the incident radiation. The IF is transmitted to a 50 Ω coaxial transmission line to a room-temperature amplifier which is connected to a spectrum analyzer. The DC biasing is room temperature filtered.

upper frequency boundary, exhibiting the maximum output power of 70 μW at 890 GHz. The BWO does not incorporate a possibility to tune the output power. The polarized radiation is coupled via free-space to the bath cryostat with the mixer block installed.

In order to achieve an ability to tune the power transmitted to the mixer block, a polarizing wire grid is installed in between the BWO and the cryostat. By adjusting the angle of the rotatable wire grid, the incident radiation intensity I_0 can be attenuated to I_1 according to Malus's law

$$I_1 = I_0 \cos^2(\theta) \tag{7.1}$$

with the angle θ between the polarization of the radiation and the polarizer. This rather simple setup is not yet optimized for ultra low-noise operation yet, but realized only to investigate on the rudimentary operation of the mixer. This manifests itself in the usage of a room-temperature amplifier, rather than a low-noise amplifier on the cold plate. The output signal is transmitted by a 50 Ω coaxial transmission line to room temperature where the amplifier is connected. The output of the amplifier is then connected to a spectrum analyzer. Similar to the amplification, the filtering of the biasing signals is not optimized for a low-noise measurement setup. Simplifying the setup, the filtering was done at room temperature, resulting in an increased noise level in the system as compared to the dipstick measurements. Since the devices under investigation exhibit a FFO to mixer coupling at frequencies lower than the designed frequency, the initial measurements were done using the BWO without the frequency tripler, thus operating in the high-power low frequency regime at a frequency of 381 GHz.

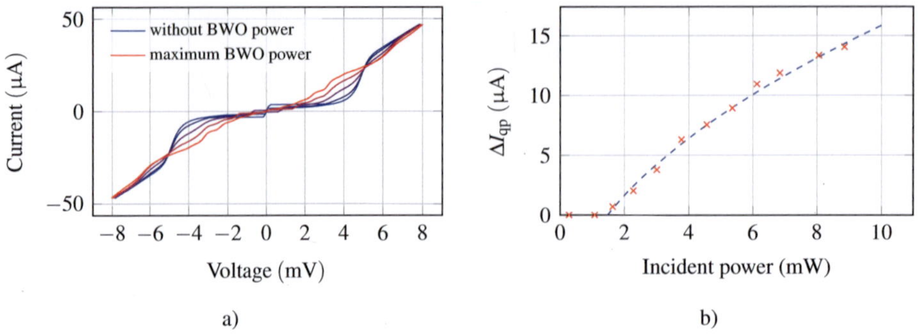

Figure 7.9: (a) Influence of the BWO 381 GHz radiation on the mixer current-voltage characteristic. (b) The measured quasiparticle current rise depending on the incident power. The dashed blue line is a square root fit.

The antenna on this device was designed to operate at the target frequency of 750 GHz. Since the fractional bandwidth of the double slot antenna rarely exceeds 20 %, most of the incident radiation is expected to be reflected by the antenna. In fact, from the simulations, it is extracted, that the reflection at this frequency is at levels as high as $|\underline{S}_{11}| = -1$ dB, thus only transmitting a small fraction of the incident power to the transmission line on chip. The result of this investigation is depicted in figure 7.9. The high power of 9.1 mW in this operation point results in a significant, even overcritical pumping level of the mixer, which is depicted in figure 7.9a. This graph contains an overlay of multiple current-voltage characteristics of the mixer, each of them recorded for a different polarizer angle, thus at a different power level. The blue curve represents the mixer characteristic with a maximally attenuated radiation ($\theta = 90°$). Constantly decreasing the polarizer angle results in an increase in power transmitted towards the mixer. This results in an increased power level at the mixer junction, yielding a higher ΔI_{qp} at a given bias voltage. For clarity reasons, only a selection of current-voltage curves is depicted in this graph. The power dependence of the quasiparticle current rise is depicted in figure 7.9b. The power given in this graph refers to the power right after the polarizer grid. Powers smaller than 1 mW do not result in a measurable quasiparticle current rise in the mixer current-voltage characteristic. Increasing from this value, the current rise follows a square root dependence on the incident power, which is expected from theory (compare equations (3.45) and (3.48)). A square root fit is given in the figure by the blue dashed line.

In the more interesting frequency range, this device did not exhibit a measurable impact from the FFO radiation on the mixer junction. Nevertheless, the device was also characterized at frequencies around 750 GHz, using the tripler in front of the BWO. The radiation

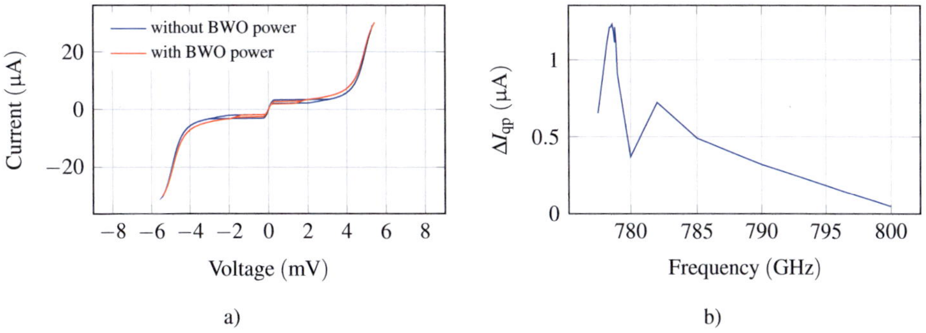

Figure 7.10: In (a), the current-voltage characteristics of the mixer junction with and without BWO radiation of 777 GHz are depicted. (b) The obtained spectrum for BWO radiation frequencies from 775 to 800 GHz.

source is thus in the low power regime. By sweeping the BWO frequency and monitoring the mixer current-voltage characteristic, a possible radiation impact on the mixer junction was searched. Figure 7.10a depicts the measured mixer characteristic without incident power. At a frequency of 777 GHz, indeed a small current rise can be observed. The current rise is however very small with only $\Delta I_{\mathrm{qp}} = 1.3\ \mu\mathrm{A}$. Nevertheless, this value is above the noise level of the measurement system and thus, a spectral dependence should be recordable. The resulting frequency dependent quasiparticle current rise is depicted in figure 7.10b. Clearly visible is a maximum in accepted RF power at a frequency of 777 GHz with a sharp drop towards lower frequencies and a moderate decrease in quasiparticle current rise towards higher frequencies. This result quite accurately correlates with the simulated operation frequency with a deviation of only about 4 %.

Nevertheless, the absolute value of the power accepted by the mixer is very small compared to the overcritical pumping by the excitation with 381 GHz. This behavior can however be explained, when considering the link budget of the transmission path from the BWO to the mixer for the two different frequencies. The main reason for the massive pump level at 381 GHz is the high power available from the BWO ($P_{381\,\mathrm{GHz}} = 9.1\ \mathrm{mW} \equiv 2.9\ \mathrm{dBm}$), which is more than two orders of magnitude higher than the power at 777 GHz ($P_{777\,\mathrm{GHz}} = 25\ \mu\mathrm{W} \equiv -16\ \mathrm{dBm}$). From the fit in figure 7.9b, the power after the polarizer for a quasiparticle current rise of $\Delta I_{\mathrm{qp}} = 1.5\ \mu\mathrm{A}$ can be estimated to $P_{\mathrm{t}} = 1.94\ \mathrm{mW} \equiv 2.9\ \mathrm{dBm}$).

After the transmission of the radiation from the BWO to free space, the Gaussian beam exhibits a strongly divergent behavior. Therefore, a setup that is optimized for a high power transmission typically contains off-axis elliptical mirrors which focus the radiation on the lens on the detector block [146].

Table 7.1: Link budget for 381 GHz and 777 GHz for $\Delta I_{qp} = 1.5$ μA

	$\nu = 381$ GHz	$\nu = 775$ GHz
P_t (dBm)	2.9	−16
Coupling efficiency (dB) [150]	−6.5	−6.5
Antenna (dB)	−15.8	−2.4
Circuit losses (dB)	−28.8	−15.2
$P_{r,\text{Simulation}}$ (dBm)	−48.2	−40
$P_{r,\text{Theory}}$ (dBm)	−56.4	−52.1

The measurement setup used in this work is however not yet optimized, no mirrors are present. This results in a strongly divergent beam pattern of the radiation emitted from the BWO. Due to the lack of focusing the radiation, a considerable amount of the available output power is expected to be lost. Nevertheless, the following consideration can be used for an approximate estimation of especially the relative power levels incident to the mixer at the two frequencies 381 GHz and 777 GHz. From free space, the radiation has to pass the cryostat window and enter the silicon lens in the detector block where the power is focused on the antenna. The losses occurring in this path are summarized as coupling efficiency and have been determined experimentally in [146, 150]. However, as opposed to those, the antenna is a double slot antenna, which is why the polarization coupling efficiency is increased from 50 % to 100 %. Furthermore, the antenna coupling efficiency is excluded and listed separately in table 7.1. In total, the coupling efficiency is accounted for with −6.5 dB.

The coupling efficiency of the antenna was estimated from simulations for the respective frequencies. Also obtained from the simulation is the transmission from the antenna to the mixer for the different frequencies with −29 dB and −3.7 dB respectively. This includes mismatch reflection losses in the entire circuit on the mixer chip, including the mismatch at the junction. Summing these values leads to the power levels accepted by the mixer. For the 381 GHz signal, it is $P_r = -48.2\ \text{dBm} \equiv 15.1\ \text{nW}$ and the 777 GHz signal arrives at the mixer with $R_r = -40.1\ \text{dBm} \equiv 97.07\ \text{nW}$. An overview on this rough estimation of the link budget is given in table 7.1. According to theory, the power is proportional to the square of the frequency [16, 84]. From the values obtained by the simulations, this relation is not fulfilled, but exhibits a difference by a factor of approximately 6. This might be due to inaccuracies in the simulations. Using the theory [16], the incident power can also be approximated by evaluating the unpumped and especially the pumped mixer current-voltage characteristics.

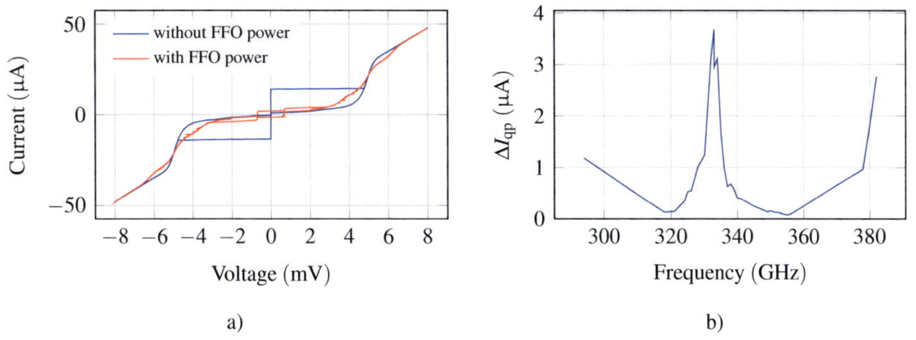

Figure 7.11: The mixer response to FFO radiation. (a) The current-voltage characteristics with and without FFO radiation at 330 GHz. (b) The spectrum of the ΔI_{qp} depending on the FFO operation frequency.

Assuming a quasiparticle current rise of $\Delta I_{\text{qp}} = 1.5\ \mu\text{A}$, this results in $P_{\text{r, Theory 381 GHz}} = 2.3\ \text{nW}$ and $P_{\text{r, Theory 777 GHz}} = 6.1\ \text{nW}$.

As expected, the power levels resulting from the simulations are approximately one order of magnitude larger than the values obtained by calculations from theory. However, as mentioned earlier, this can be explained by the non-optimal optical setup, which will be required to be improved in the future. Nevertheless, from this estimation the power coupling at the 381 GHz can be explained. In spite of the low coupling efficiency of the antenna and the tuning circuit, the BWO still offers enough power to overcritically pump the mixer.

The identical device is also characterized regarding the flux-flow operation. The results obtained by this characterization is depicted in figure 7.11. The left graph contains an exemplary current-voltage characteristic, pumped with an signal frequency of 334 GHz. Clearly visible is the quasiparticle current rise in the retrapping branch. Furthermore, the right graph depicts the spectral dependence of ΔI_{qp}, revealing a radiation power acceptance by the mixer at frequencies of 334 and 380 GHz. However, no spectral information is available for frequencies above these values, since no ΔI_{qp} was measurable.

In order to compare the results from the FFO operation with the simulation, the spectra are depicted in a single graph in figure 7.12. The red line represents the simulated transmission characteristic of the FFO to the mixer junction. The measured transmission peaks correspond well with the simulated ones at $\nu = 334$ and 380 GHz. However, it is unclear, why only these peaks can be measured, whereas the peaks towards higher frequencies with even higher simulated transmission values are not measurable. In addition to the FFO operation related curves, the graph also contains the transmission simulated from the antenna to the mixer as the green curve. This can be compared with the results shown earlier obtained from incident power from the BWO which is depicted by the blue line.

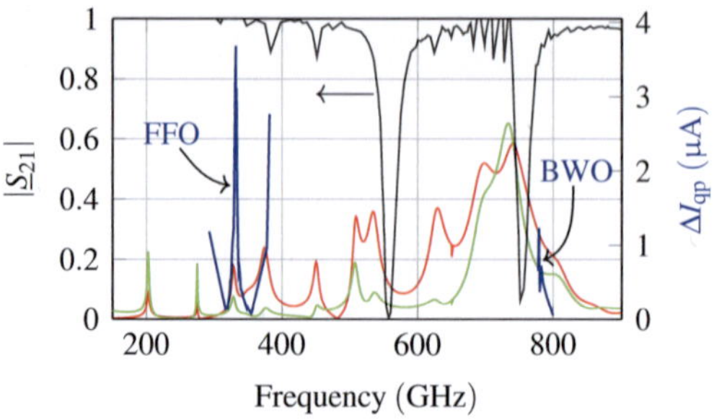

Figure 7.12: Comparison of the simulated transmission spectra with the measured ones. The green and red line represent the simulated transmission from the antenna and the FFO to the mixer. The blue lines depicts the measured quasiparticle current rise induced by the FFO and the BWO. The black line shows the atmospheric transmission, adopted from [151].

As predicted by the simulation, the mixer accepts power from the BWO at frequencies at 777 GHz. The sudden decay in transmitted power can be explained by the atmospheric transmission. The absorption maximum of the atmosphere at frequencies around 750 GHz further suppresses the already small values of the incident power, resulting in a non-measurable response of the mixer to the BWO radiation. The spectral response of the mixer has not yet been measured for the frequency range below the atmospherical absorption maximum.

The combined results from the FFO operation and the BWO generated free-space radiation lead to the assumption, that the mixer tuning simulation is valid. However, the fact that no power from the FFO can be measured at frequencies higher than 380 GHz can not be explained by these simulations. Since the mixer tuning is assumed valid, the error has to be located in the path between FFO and the T-junction. In order to further validate the simulation, the spectral response to free-space radiation has to be investigated more closely. This includes the optimization of the measurement setup to maximize the power levels at the mixer. Furthermore, the setup needs to be optimized regarding the noise level that is present at the mixer, due to the lack of proper filtering. This becomes especially important, when the receivers are to be characterized in the future regarding their noise temperature. With an optimized measurement setup, the spectral response of the receivers can be obtained with a larger resolution, enabling the identification of smaller intensity transmission peaks as compared to the current setup. Furthermore, it is required to record the receiver response across a larger spectral range, enabling the identification of the different transmission peaks. This will facilitate the validation of the simulations.

Another approach, to locate possible reasons for the lack of the high-frequency operation, was conducted using a different mixer installed in the cryostat. This device was realized using a SIS junction with an area of $0.9 \times 0.9\ \mu m^2$, resulting in a normal state resistance reduced

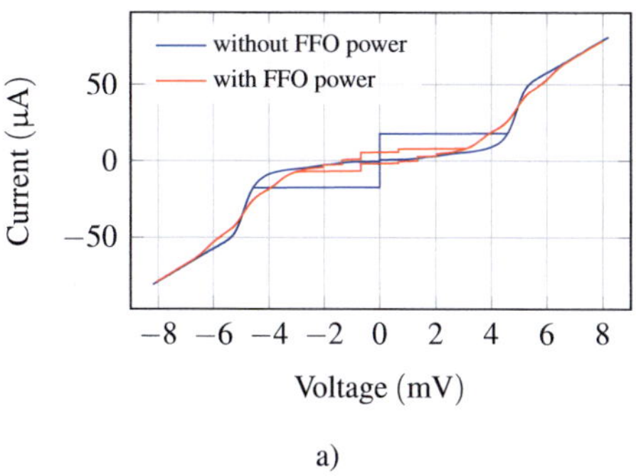

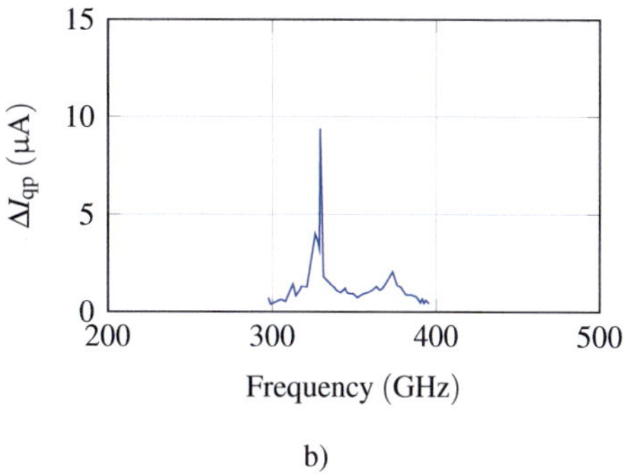

Figure 7.13: (a) Current-voltage characteristic of a mixer with increased area (0.9×0.9 µm). (b) The measured spectral dependence of the power transmitted from the FFO to the mixer.

by 30 % compared to the previous devices ($R_N = 95\ \Omega$). The current-voltage characteristic of this device is shown in figure 7.13a with and without applied FFO power at 330 GHz. The operation frequency of the coupling is again in the range of 300 − 400 GHz, as can be seen in the spectrum of the quasiparticle current rise, depicted in figure 7.13b. Apparently, the lower normal state resistance has no significant influence on the operation frequency of the mixer, when driven with the FFO, since no mixer response to a high frequency input radiation was measurable.

As the device shown previously, also this one exhibits a significant power coupling of the FFO at the low frequency regime, in which the BWO exhibits a high output power level. Therefore, this device is used for preliminary investigations on its mixing properties, demonstrating the basic mixing operation. As opposed to the regular operation, the FFO in this case serves as the source of the RF signal instead of the LO signal, which is provided by the BWO radiation at 330 GHz. The intermediate frequency output is connected to an amplifier at room temperature through stainless steel coaxial transmission lines. These transmission lines are installed in spite of their rather high attenuation due to their low thermal load to the cryostat. However, since the amplifier is installed only at room temperature instead of right behind the mixer block, the output power of the mixer is significantly attenuated before even reaching the amplifier. Figure 7.14 depicts an IF output spectrum recorded with a spectrum analyzer. The black curve depicts the noise level of the system at around −72 dBm. The intermediate frequency output spectrum around 2 GHz demonstrates the basic operation of the mixing process. This measurement is not suitable for e.g. the determination of the linewidth of the FFO, since the bandwidth of the BWO is unknown, and only specified with approximately 3 % in the data sheet provided by the manufacturer.

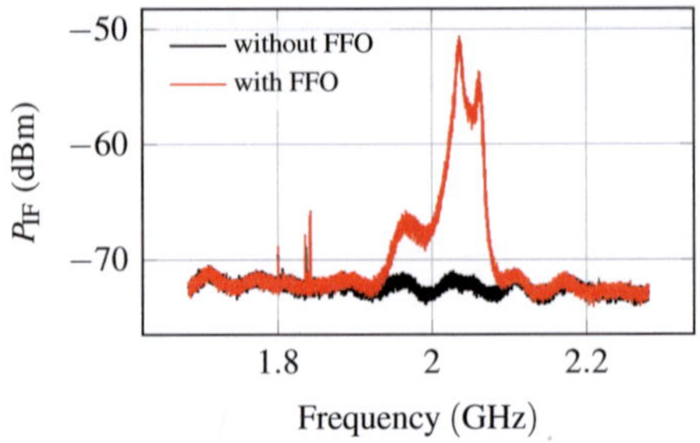

Figure 7.14: Intermediate frequency output of the integrated receiver for incident 330 GHz radiation. The BWO is used as local oscillator, the FFO is used as signal source. The two curves are recorded with and without applied FFO power respectively.

Therefore, it can not be determined whether the bandwidth in the IF spectrum stems from the BWO radiation or results from the flux-flow oscillator. Nevertheless, this preliminary result shows the successful mixing of a radiation signal received from free space with a radiation signal generated on chip by a NbN-based flux-flow oscillator.

In order to characterize the mixer regarding the ideal bias point for a maximized conversion, the amplitude of the intermediate frequency output has to be monitored at different mixer bias points. This can be done with the current setup by setting the spectrum analyzer to a zero span mode which then monitors a specific frequency in time domain. The spectrum analyzer contains an analogue output, which can be connected to a lock-in amplifier, in order to quantify the output power measured by the spectrum analyzer. By sweeping the bias voltage of the mixer, and simultaneous recording of the lock-in amplifier signal, the dependence of the mixer conversion efficiency on the bias point can be obtained. The result of this measurement is shown in figure 7.15. The blue curve depicts the unpumped current-voltage characteristic of the mixer as a reference. The red curve is obtained for this mixer, pumped with the BWO at $\nu = 330\,\text{GHz}$. The black curve depicts the recorded lock-in signal, where the BWO radiation was chopped. Starting from zero bias voltage, an increase in output power can be seen, which peaks on the first photon step. For voltages exceeding the gap voltage, the measured output power drops rapidly. In literature [84] it is found that low LO power levels typically result in a maximized conversion efficiency of the mixer at the first photon step. This is exactly the behavior that is shown by the present receiver.

With these fully integrated NbN-based receivers, a successful mixer operation was demonstrated, yielding an intermediate frequency spectrum from RF frequencies at 330 GHz. Since the devices were designed for 750 GHz, the coupling efficiency for free space radiation is not at the required high values. Therefore, the receivers were not characterized regarding their

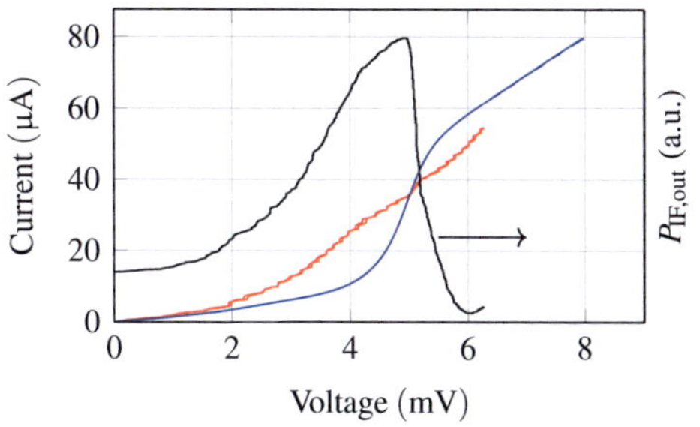

Figure 7.15: Measurement of the IF output power of the mixer, depending on the DC bias point. The blue curve depicts the mixer current-voltage characteristic without external radiation. By applying BWO radiation, the mixer is pumped (red curve). The mixer bias point is swept and the analog output of the spectrum analyzer is connected to a lock-in amplifier.

noise temperature. Nevertheless, these devices prove the successful implementation of an integrated receiver in niobium nitride technology. Several receiver devices have been fabricated where all of them exhibited the lack of efficient radiation transmission at the targeted frequency. A preliminary parameter variation did not yield the origin of this behavior. By optimizing the spectral coupling efficiency of the FFO to the mixer at the target frequency, an efficient pumping of the mixer by the flux-flow oscillator can be achieved. This will enable the low-noise operation of the NbN-based integrated receiver at frequencies exceeding the 700 GHz limit.

Conclusion

Multiple integrated receivers were characterized by quasioptical measurements in the cryostat regarding various characteristics of the devices under test. The measurements setup comprised a backward wave-oscillator as radiation source, irradiating the mixer through the cryostat window. This preliminary measurement setup is not yet optimized for the accurate measurements of the sensitivity of the receivers. Nevertheless, a significant amount of radiation was recorded at the mixer for a frequency of 381 GHz due to the high power of the BWO at this frequency. Although the FFO operation did not result in a measurable mixer pumping, the irradiation of the mixer by the BWO with a frequency of 777 GHz resulted in a small, but measurable quasiparticle current rise. Considering the low power levels of the BWO at this frequency, this small ΔI_{qp} value of the mixer response to this signal is in a reasonable range. This result leads to the assumption that the signal path from the antenna to the mixer tuning behaves close to the predictions from the simulation. Due the fact that no impact from the FFO radiation was measurable at these frequencies, the error in simulation is expected to occur in the signal path from the FFO to the mixer. Since the BWO offers high power at the low frequency regime, the receiver operation was demonstrated at a frequency of 330 GHz, yielding an intermediate frequency signal from the FFO and the BWO radiation. The dependence of the IF output power on the mixer bias point was investigated, yielding the expected result of the highest output power for a mixer biasing on the first photon step. Although the receiver operation was not demonstrated at the desired target frequency, these measurements prove the successful implementation of an integrated receiver in NbN/AlN/NbN technology.

8 Summary

Superconducting integrated THz receivers comprise a flux-flow oscillator integrated on chip with a SIS mixer and a receiving antenna. These two active components and the antenna are interconnected by a transmission line network, which has to ensure a low loss transmission of the radiation between the three components. This circuit furthermore has to allow a separate DC biasing of the active components and the tuning of the mixer reactance. An accurate modeling of this complex RF-network can be achieved only with the knowledge of the material parameters of the applied superconductors and dielectrics. The most common superconductor is Nb, which limits the operation frequency to values of $\leq 700\,\mathrm{GHz}$. Niobium nitride exhibits an energy gap twice as large as niobium, which is why the substitution of Nb by NbN as electrode material allows an increase in operation frequency up to 1.4 THz. The frequency limitation includes the mixer junction itself, but also and even more significantly, the flux-flow oscillator and the transmission lines on chip. In addition to the large energy gap, a high-quality receiver also requires ultra-low leakage currents and simultaneously large critical current densities of the mixer junctions, which requires the surface of the ground electrode to be smooth on a sub-nm scale. As known from literature and as shown by experiments during this work, the highest quality of NbN-based Josephson junctions is achieved on magnesium oxide substrates. However, due to the hygroscopic nature of MgO it is beneficial to realize the trilayers on more resistant substrates on which a high quality NbN deposition poses a significant challenge.

Therefore, this work was dedicated to the realization of a superconducting integrated receiver with niobium nitride electrodes on substrates other than MgO. This requires a close optimization of the deposition process along with the development of a reliable fabrication technology. The deposition process was investigated and optimized for multiple substrate materials such as oxidized silicon, silicon and sapphire. Due to the complexity of the reactive deposition process, the various deposition parameters are highly correlated. The critical temperature can be optimized using the nitrogen consumption as a parameter. It was found that the total pressure for a constant argon to nitrogen ratio and a constant relative nitrogen consumption only exhibits a minor influence on the transition temperature of the NbN films. The surface roughness however, is strongly determined by the total pressure. Using these correlations, the independent optimization of the two most important quality parameters of NbN films is enabled. The obtained results are a significant improvement over the initial optimization point, yielding NbN films with critical temperatures of 15.7 K and a RMS surface roughness of less than 1 nm at a film thickness of 200 nm on sapphire substrates. The maximization of the critical temperature yields junctions with gap voltages as high as 5.1 mV, which should enable the operation of the superconducting integrated receiver well above the 1 THz limit. The low surface roughness of the ground electrode results in low leakage currents at low critical current densities. Increasing the critical current density is done by a variation of the deposition time of the AlN barrier, resulting in thinner barriers. This leads to a strong increase in leakage currents, which is why the surface roughness will have to be improved further, allowing for low leakage currents through the barrier at higher critical current densities. In general, the deposition system will have to be further optimized, in order to decrease the spread of the quality parameter of the resulting trilayers. Furthermore, the properties of the wiring layer need to be reconsidered. Since the NbN wiring layer is deposited on amorphous surfaces, it exhibits a significantly lower energy gap than the electrodes. In order to achieve the highest possible low loss operation frequency, the energy gap of the wiring layer needs to be at the same high level as the electrode energy gap.

For the realization of an integrated receiver, the patterning is of utmost importance, since the characteristic impedances of the transmission lines are dependent on the lateral dimensions of the structures. Therefore, the patterning was optimized, resulting in achievable mixer junction areas down to 0.5×0.5 μm^2 and minimum linewidths of the wiring layer as small as 500 nm. The microstrip transmission lines are created by the wiring layer, separated from the NbN ground electrode by a 400 nm thick SiO dielectric. This material setup leads to a characteristic impedance range from 70 Ω down to 10 Ω, which enables the impedance matching between the FFO, the mixer and the antenna. For the proper design of the RF circuits, the knowledge of the surface impedances of the employed materials is required. These values are extracted from temperature dependent resistance measurements, subsequent calculation using the Mattis-Bardeen theory and can then be included in the simulation software CST microwave studio. Utilizing this software, RF circuits were developed for the realization of an integrated receiver. Due to the large complexity of the integrated receiver, prototype devices were designed, realized and characterized, in order to verify the simulation results. These prototypes were created with a consecutive increase in complexity, starting from a mixer integrated with an antenna for 166 GHz in niobium nitride technology, without a flux-flow oscillator. The subsequent prototypes disregarded the antenna but included the flux-flow oscillator instead, with the more complex microstrip network between the FFO and the mixer. This transmission line network needs to allow for a parallel biasing of the FFO and the mixer, hence disconnect both components for DC currents, but on the other hand grant a maximized transmission of the RF signal. Furthermore, the creation of localized magnetic fields needs to be included, since both, the mixer as well as the FFO require magnetic fields of different strengths for operation. This is achieved by the implementation of control lines, which create the localized magnetic fields by a current parallel to the electrodes in the vicinity of the mixer and the flux-flow oscillator. The characterization of these preliminary devices resulted in a demonstration of radiation transmission from the FFO to the mixer for frequencies as high as 600 GHz, agreeing well with the simulated values. Furthermore, multiple simulation challenges could be verified. This included the verification of the material parameters, as well as the correct representation of the active devices in the circuit. The operation of the preliminary devices was limited to frequencies around 400 GHz and 600 GHz, which is not imposed by the flux-flow oscillator or the microstrip lines, but is the design frequency, where the mixer capacity is compensated by the tuning circuit.

Different tuning approaches are introduced, realized and characterized: the classical stub tuning which simply connects an inductor in parallel to the mixer capacity, the twin junction tuning approach, which uses a second junction in parallel to the actual mixer junction and the resulting total capacity is compensated by a inductor in between these junctions. The third tuning approach is the hybrid end-loaded stub tuning, which has the same setup as the twin junction tuning, except, the second capacity is realized not by a second junction, but by a capacitor. The realization of these three tuning approaches allowed their characterization regarding the frequency accuracy and frequency tuning capability. This leads to the observation of the best tuning structure in terms of accuracy being the twin junction tuning, which exhibited an almost perfect match between the simulated and the measured transmission peaks. The hybrid end-loaded stub tuning also exhibited a good agreement between simulation and measurement, however, with a slightly increased deviation, followed by the classical stub tuning with an even higher deviation. All three approaches could thus be demonstrated to operate, with a varying frequency accuracy. Furthermore, the upscaling of the frequency exhibits different limits for the different tuning approaches. All inductors are realized using transmission lines, which exhibit a length dependent inductance. Higher operation frequencies thus lead to a shorter length of this transmission line for a given mixer capacity. The classical tuning approach exhibits the shortest transmission line length among the three investigated approaches. Therefore, the operation frequency is limited in practice due to calculated tuning lengths in the deep sub-micrometer range for high frequencies. This tuning length is in this case already within the accuracy limit of the patterning, hence the frequency accuracy is deteriorated. Due to increased tuning lengths in the twin junction tuning and the end-loaded stub tuning approach, these approaches can be used for higher frequency operation of the mixer.

The results of the prototype devices were included in the design of a fully integrated receiver, which is the next step in increasing the total complexity of the RF circuit. The included double-slot antenna is connected to a T-junction along with the flux-flow oscillator. The signals coming from these two components are transmitted to the mixer, which is embedded in a hybrid end-loaded stub tuning network. This tuning network was selected, since it offers the possibility of realizing the control line in the ground electrode. The resulting devices proved to work with significantly higher pumping levels compared to the previous prototypes, however, the operation frequency was significantly below the design frequency. Multiple possible origins of these deviations have been considered as the reason for this behavior, such as the inaccuracy in patterning at the intersection of the butterfly stub and the microstrip line, a deviating FFO output impedance, a mismatch induced by the T-junction, or an additional imaginary part in impedance occurring at the mixer site, due to either deviating mixer capacities or non-zero input imaginary impedance stemming from the RF choke at the intermediate frequency output. However, none of these have resulted in a satisfactory explanation for the measured behavior. Nevertheless, a direct detection of incident BWO radiation was possible for frequencies around 777 GHz. A receiver operation was demonstrated for frequencies around 330 GHz using the BWO in conjunction with the flux-flow oscillator. This included the recording of an intermediate frequency spectrum as well as the bias point induced variation in IF output power of the mixer. The operation of an integrated receiver with a device based on NbN/AlN/NbN technology was thus demonstrated with these devices. A close investigation of the operation frequency deviation of the presented devices should enable the reliable realization of integrated receivers at frequencies above 1 THz.

Nomenclature

α	Attenuation constant of an electromagnetic wave in a transmission line $\left(\frac{1}{\mathrm{m}}\right)$
α_{SG}	Damping parameter in the sine-Gordon equation ()
α_{SIS}	Pumping parameter of a SIS mixer ()
α_{Stub}	Opening angle of a radial stub (deg)
$\bar{c}$	Swihart velocity $\left(\frac{\mathrm{m}}{\mathrm{s}}\right)$
β	Imaginary part of the propagation constant γ of an electromagnetic wave in a transmission line $\left(\frac{1}{\mathrm{m}}\right)$
β_{C}	Steward-McCumber parameter (V)
β_{SG}	Surface resistance of the superconductors in the Sine-Gordon equation ()
ΔI_{qp}	Quasiparticle current rise in the subgap branch of a SIS mixer (A)
Δp_{N2}	Nitrogen consumption during a reactive deposition process (%)
Δ	Single energy gap of a superconductor (eV)
δ_{s}	Skin depth (m)
ℓ	Length of a transmission line (m)
ε_0	Vacuum permittivity $\left(8.854\ 10^{-12}\ \frac{\mathrm{F}}{\mathrm{m}}\right)$
ε_{r}	Relative permittivity ()
η_{F}	Friction coefficient in the RCSJ model ()
Γ	Reflection coefficient ()
γ	Propagation constant of an electromagnetic wave $\alpha + \mathrm{j}\beta$ $\left(\frac{1}{\mathrm{m}}\right)$
$\hbar$	Reduced Planck constant $\frac{h}{2\pi}$
κ	Ginzburg-Landau parameter ()
λ_{J}	Josephson penetration depth (m)
λ_{L}	London penetration depth of a superconductor (m)

μ_0	Vacuum permeability $\left(12.566\ 10^{-7}\ \frac{\mathrm{H}}{\mathrm{m}}\right)$
ν	Frequency (Hz)
ν_{LO}	Local oscillator frequency (Hz)
ν_{RF}	Incident signal of interest (Hz)
ω	Angular frequency $2\pi\nu$ $\left(\frac{1}{\mathrm{s}}\right)$
ω_{Plasma}	Plasma frequency in a Josephson junction $\left(\frac{1}{\mathrm{s}}\right)$
ω_{res}	Resonance angular frequency of a resonant circuit $\left(\frac{1}{\mathrm{s}}\right)$
Φ_0	Magnetic flux quantum $\frac{\mathrm{h}}{2\mathrm{e}} = 2.0678 \cdot 10^{-15}$ Wb
Ψ	Superconducting wavefunction
ρ	Specific resistance (Ωm)
ρ_{20}	Specific resistance at $T = 20$ K (Ωm)
σ	Specific conductance $\left(\frac{\mathrm{S}}{\mathrm{m}}\right)$
σ_{n}	Normal state conductance of a superconductor $\left(\frac{\mathrm{S}}{\mathrm{m}}\right)$
$\underline{S}_{\mathrm{i,j}}$	Scattering parameter for port i and j (dB)
$\underline{Z}_{\mathrm{emb}}$	Embedding impedance of a mixer junction (Ω)
φ	Phase difference of two superconductors across a Josephson junction
φ_{i}	Phase of a superconductor
$\vec{B}$	Magnetic flux density (T)
$\vec{E}$	Electrical field strength $\left(\frac{\mathrm{V}}{\mathrm{m}}\right)$
ξ	BCS-coherence length (m)
ξ_{GL}	Ginzburg- Landau coherence length (m)
ζ	Specific resistance of the superconductor (Ω)
ζ_{S}	Specific surface resistance of a superconductor (Ω)
$B_{\mathrm{c,2}}$	Second critical magnetic field (T)
C	Capacity (F)
C'	Specific capacity $\left(\frac{\mathrm{F}}{\mathrm{m}}\right)$
C'_{tri}	Capacitance of a SIS trilayer $\left(\frac{\mathrm{F}}{\mathrm{m}}\right)$
d_{CPW}	Total width of the coplanar waveguide (m)
d_{diel}	Dielectric thickness between a microstrip and the ground electrode (m)
e	Elementary charge ($1.602 \cdot 10^{-19}$ C)

G	Conductivity (S)
G_{amp}	Gain of an amplifier ()
h	Planck constant (6.626 10^{-34} Js)
h_1	Width of the active part of the output of a flux-flow oscillator (m)
h_2	Width of the idle region of the FFO output (m)
H_{c}	Critical magnetic field of a superconductor $\left(\frac{\mathrm{A}}{\mathrm{m}}\right)$
I_{bias}	Bias current of a mixer or a flux-flow oscillator (A)
I_{CTRL}	Control line current (A)
I_{c}	Critical Josephson current (A)
$I_{\mathrm{c}}R_{\mathrm{N}}$	Ambegaokar-Baratoff parameter (V)
I_{dis}	Discharge current of the plasma in the magnetron sputtering device (A)
$I_{\mathrm{Q,RCSJ}}$	Quasiparticle current in the RCSJ model (A)
I_{S}	Superconducting current through a Josephson junction (A)
I_{total}	Total current through a junction in the RCSJ model (A)
$I_{\mathrm{V,RCSJ}}$	Displacement current in the RCSJ model (A)
$j_{\mathrm{c,SC}}$	Critical current density of a superconductor $\left(\frac{\mathrm{A}}{\mathrm{cm}^2}\right)$
j_{c}	Critical current density of a Josephson junction $\left(\frac{\mathrm{A}}{\mathrm{cm}^2}\right)$
k_{B}	Boltzmann-constant ($1.38 \cdot 10^{-23}$ m^2 kg s^{-2} K^{-1})
L	Inductance (H)
L'	Specific inductance $\left(\frac{\mathrm{H}}{\mathrm{m}}\right)$
L_{JJ}	Length of a Josephson junction (m)
l_{JJ}	Length of a Josephson junction, normalized to the Josephson penetration depth ()
m_{S}	Mass of Cooper pairs (kg)
n_{S}	Density of the Cooper pairs ()
p_{Ar}	Argon pressure in the recipient (bar)
P_{IF}	Intermediate frequency power generated by a mixer (W)
P_{LO}	Local oscillator power available at the mixer junction (W)
P_{L}	Power dissipated in a load R_{L} (W)
p_{N2}	Nitrogen pressure in the recipient (bar)

P_n	Noise power (W)
P_{RF}	Overall power of a signal of interest (W)
q	Charge of a Cooper pair (C)
R	Resistance (Ω)
$R_{d,bias}$	Differential resistance of a flux-flow oscillator derivated from the bias current (Ω)
$R_{d,CTRL}$	Differential resistance of a FFO derivated from the control line current (Ω)
R_{mix}	Mixer normal state resistance (Ω)
R_N	Normal state resistance of a Josephson junction (Ω)
R_{sg}	Resistance in the subgap branch of a Josephson junction (Ω)
RRR	Residual resistance ratio ()
s	Gap width in a coplanar waveguide (m)
t	Time (s)
t_1	Thickness of the ground electrode of a SIS junction (m)
t_2	Thickness of the counter electrode of a SIS junction (m)
t_3	Thickness of the wiring layer (m)
T_c	Critical temperature of a superconductor (K)
t_{diel}	Thickness of the dielectric between the wiring layer and the ground electrode (m)
T_n	Noise temperature (K)
$t_{ox,m}$	Magnetic thickness of the barrier in a SIS junction (m)
t_{ox}	Physical thickness of the barrier in a SIS junction (m)
V_0	Mixer bias voltage (V)
V_{dis}	Discharge voltage of the plasma in the DC magnetron sputtering device (V)
v_f	Fermi velocity $\left(\frac{m}{s}\right)$
V_{gap}	Gap voltage of a Josephson junction (V)
V_{JJ}	Voltage across a Josephson junction (V)
V_{LO}	Voltage of the local oscillator (V)

V_{PAT}	Voltage span of a photon-assisted tunneling step (V)
v_{p}	Phase velocity $\left(\frac{\mathrm{m}}{\mathrm{s}}\right)$
v_{p}	Phase velocity of an electromagnetic wave $\left(\frac{\mathrm{m}}{\mathrm{s}}\right)$
V_{Shap}	Voltage span of a Shapiro step (V)
w_{CPW}	Width of the center conductor of a coplanar waveguide (m)
W_{JJ}	Width of a Josephson junction (m)
w_{JJ}	Width of a Josephson junction, normalized to the Josephson penetration depth ()
w_{MS}	Width of a microstrip line (m)
X_{mix}	Mixer reactance (Ω)
z	Spatial coordinate in the propagation direction of an electromagnetic wave (m)
Z_0	Characteristic impedance of a transmission line (Ω)
Z_{FFO}	FFO output impedance (Ω)
Z_{g}	Generator output impedance (Ω)
Z_{L}	Load impedance (Ω)
$Z_{\mathrm{S,N}}$	Surface impedance of a normal conductor (Ω)
Z_{S}	Surface impedance of a superconductor (Ω)
AC	Alternating current
AFM	Atomic-force microscope
BW	Bandwidth (Hz)
BWO	Backward wave oscillator
CPW	Coplanar waveguide
DC	Direct current
DUT	Device under test
FFO	Flux-flow oscillator
h-ELS	Hybrid end-loaded stub
HDPE	High-density polyethylene
IF	Intermediate frequency
JJ	Josephson junction

LNA	Low noise amplifier
MS	Microstrip transmission line
RCSJ	Resistively and Capacitively Shunted Junction
RF	Radio Frequency
SEM	Scanning electron microscope
SIS	Superconductor Insulator Superconductor
twinJJ	Twin junction

List of Figures

List of Tables

Bibliography

[1] J. W. Kooi, *Advanced receivers for submillimeter and far infrared astronomy*. University of Groningen, 2008.

[2] K. Gundlach and M. Schicke, "SIS and bolometer mixers for terahertz frequencies," *Superconductor Science and Technology*, vol. 13, no. 12, p. R171, 2000.

[3] V. P. Koshelets, A. B. Ermakov, L. V. Filippenko, N. V. Kinev, O. S. Kiselev, M. Y. Torgashin, A. De Lange, G. De Lange, S. I. Pripolzin, and V. L. Vaks, "Superconducting integrated THz receivers: development and applications," in *Infrared, Millimeter Wave, and Terahertz Technologies*, vol. 7854, p. 78540J, International Society for Optics and Photonics, 2010.

[4] T. G. Phillips and J. Keene, "Submillimeter astronomy (heterodyne spectroscopy)," *Proceedings of the IEEE*, vol. 80, no. 11, pp. 1662–1678, 1992.

[5] A. Lecacheux, N. Biver, J. Crovisier, D. Bockelée-Morvan, P. Baron, R. Booth, P. Encrenaz, H.-G. Florén, U. Frisk, Å. Hjalmarson, *et al.*, "Observations of water in comets with Odin," *Astronomy & Astrophysics*, vol. 402, no. 3, pp. L55–L58, 2003.

[6] P. Schilke, T. Groesbeck, G. Blake, and T. Phillips, "A line survey of orion KL from 325 to 360 GHz," *The Astrophysical Journal Supplement Series*, vol. 108, no. 1, p. 301, 1997.

[7] P. Schilke, D. Benford, T. Hunter, D. Lis, and T. Phillips, "A line survey of Orion-KL from 607 to 725 GHz," *The Astrophysical Journal Supplement Series*, vol. 132, no. 2, p. 281, 2001.

[8] A. Krabbe and S. C. Casey, "First-light SOFIA instruments," in *Infrared Spaceborne Remote Sensing X*, vol. 4818, pp. 1–14, International Society for Optics and Photonics, 2002.

[9] A. Kerr, S.-K. Pan, and W. Lyons, "The genesis of SIS mixers-the legacy of John Tucker in radio astronomy," in *Microwave Symposium (IMS), 2015 IEEE MTT-S International*, pp. 1–4, IEEE, 2015.

[10] T. Phillips, D. Woody, G. Dolan, R. Miller, and R. Linke, "Dayem-Martin (SIS tunnel junction) mixers for low noise heterodyne receivers," *IEEE Transactions on Magnetics*, vol. 17, no. 1, pp. 684–689, 1981.

[11] A. M. Barychev, *Superconductor-insulator-superconductor THz mixer integrated with a superconducting flux-flow oscillator*. University of Groningen, 2005.

[12] K. Il'in, A. Semenov, H.-W. Hubers, and M. Siegel, "Hot-electron bolometer mixers for terahertz radiation," *Electronics Letters*, vol. 46, no. 26, pp. s14–s16, 2010.

[13] J. J. Baselmans, M. Hajenius, J. Gao, T. Klapwijk, P. De Korte, B. Voronov, and G. Gol'tsman, "Doubling of sensitivity and bandwidth in phonon cooled hot electron bolometer mixers," *Applied physics letters*, vol. 84, no. 11, pp. 1958–1960, 2004.

[14] D. Cunnane, J. H. Kawamura, M. A. Wolak, N. Acharya, X. X. Xi, and B. S. Karasik, "Optimization of parameters of MgB 2 hot-electron bolometers," *IEEE Transactions on Applied Superconductivity*, vol. 27, no. 4, pp. 1–5, 2017.

[15] J. Tucker, "Quantum limited detection in tunnel junction mixers," *IEEE Journal of Quantum Electronics*, vol. 15, no. 11, pp. 1234–1258, 1979.

[16] J. R. Tucker and M. J. Feldman, "Quantum detection at millimeter wavelengths," *Reviews of Modern Physics*, vol. 57, no. 4, p. 1055, 1985.

[17] A. Khudchenko, A. M. Baryshev, K. I. Rudakov, P. M. Dmitriev, R. Hesper, L. de Jong, and V. P. Koshelets, "High-gap Nb-AlN-NbN SIS junctions for frequency band 790–950 GHz," *IEEE Transactions on Terahertz Science and Technology*, vol. 6, no. 1, pp. 127–132, 2016.

[18] M. Yao, J. Li, D. Liu, S. Li, S. Shi, P. Dmitriev, M. Fominsky, and V. Koshlets, "Development of a 0.85 THz Nb-AlN-NbN superconductor-insulator-superconductor mixer," in *Millimeter-Waves, 2017 10th Global Symposium on*, pp. 117–119, IEEE, 2017.

[19] A. Karpov, D. Miller, F. Rice, J. Stern, B. Bumble, H. LeDuc, and J. Zmuidzinas, "Low Noise 1 THz–1.4 THz Mixers Using Nb/Al-AlN/NbTiN SIS Junctions," *IEEE Transactions on Applied Superconductivity*, vol. 17, no. 2, pp. 343–346, 2007.

[20] A. Karpov, D. Miller, F. Rice, J. Stern, B. Bumble, H. LeDuc, and J. Zmuidzinas, "Development of 1.25 THz SIS mixer for Herschel space observatory," in *Millimeter and Submillimeter Detectors and Instrumentation for Astronomy III*, vol. 6275, p. 62751U, International Society for Optics and Photonics, 2006.

[21] A. Karpov, D. A. Miller, J. A. Stern, B. Bumble, H. G. LeDuc, I. Mehdi, R. H. Lin, and J. Zmuidzinas, "Low noise 1 thz sis mixer for stratospheric observatory: Design and characterization," *IEEE Transactions on Applied Superconductivity*, vol. 21, no. 3, pp. 616–619, 2011.

[22] Y. Uzawa, Z. Wang, and A. Kawakami, "Terahertz NbN/AlN/NbN mixers with Al/SiO/NbN microstrip tuning circuits," *Applied physics letters*, vol. 73, no. 5, pp. 680–682, 1998.

[23] M. L. Edgar and J. Zmuidzinas, "CASIMIR: A submillimeter heterodyne spectrometer for SOFIA," in *Airborne Telescope Systems*, vol. 4014, pp. 31–43, International Society for Optics and Photonics, 2000.

[24] L. R. D'Addario, "An SIS mixer for 90–120 GHz with gain and wide bandwidth," *International journal of infrared and millimeter waves*, vol. 5, no. 11, pp. 1419–1442, 1984.

[25] J. Kooi, M. Chan, B. Bumble, H. LeDuc, P. Schaffer, and T. Phillips, "180–425 GHz low noise SIS waveguide receivers employing tuned Nb/AIO x/Nb tunnel junctions," *International journal of infrared and millimeter waves*, vol. 15, no. 5, pp. 783–805, 1994.

[26] C. Walker, J. Kooi, M. Chant, H. G. Leduc, P. Schaffer, J. Carlstrom, and T. Phillips, "A low-noise 492 GHz SIS waveguide receiver," *International journal of infrared and millimeter waves*, vol. 13, no. 6, pp. 785–798, 1992.

[27] W. Shan, S. Shi, T. Matsunaga, M. Takizawa, A. Endo, T. Noguchi, and Y. Uzawa, "Design and development of SIS mixers for ALMA band 10," *IEEE Transactions on Applied Superconductivity*, vol. 17, no. 2, pp. 363–366, 2007.

[28] J. Kawamura, J. Chen, D. Miller, J. Kooi, J. Zmuidzinas, B. Bumble, H. G. LeDuc, and J. A. Stern, "Low-noise submillimeter-wave NbTiN superconducting tunnel junction mixers," *Applied Physics Letters*, vol. 75, no. 25, pp. 4013–4015, 1999.

[29] M. Justen, M. Schultz, T. Tils, R. Teipen, S. Glenz, P. Putz, C. Honingh, and K. Jacobs, "SIS flight mixers for band 2 of the HIFI instrument of the Herschel Space Observatory," in *Infrared and Millimeter Waves, 2004 and 12th International Conference on Terahertz Electronics, 2004. Conference Digest of the 2004 Joint 29th International Conference on*, pp. 437–438, IEEE, 2004.

[30] A. Karpov, D. Miller, F. Rice, J. Stem, and B. Bumble, "Low noise NbTiN 1.25 THz SIS mixer for Herschel Space Observatory," in *Proc. 16th Int. Symp. Space Terahertz Technol.*, p. 450, 2005.

[31] S. Asayama, S. Kawashima, H. Iwashita, T. Takahashi, M. Inata, Y. Obuchi, T. Suzuki, and T. Wada, "Design and development of ALMA band 4 cartridge receiver," in *Proc. 19th Intl. Symp. Space Terahertz Tech.*, pp. 244–249, 2008.

[32] B. Billade, O. Nystrom, D. Meledin, E. Sundin, I. Lapkin, M. Fredrixon, V. Desmaris, H. Rashid, M. Strandberg, S.-E. Ferm, *et al.*, "Performance of the first ALMA band 5 production cartridge," *IEEE Transactions on Terahertz Science and Technology*, vol. 2, no. 2, pp. 208–214, 2012.

[33] J. Zmuidzinas, A. Karpov, D. Miller, F. Rice, H. G. Leduc, J. Pearson, and J. A. Stern, "Coherent detection and SIS mixers," in *Far-IR, Sub-mm and MM Detector Technology Workshop*, 2002.

[34] S. Mahieu, D. Maier, B. Lazareff, A. Navarrini, G. Celestin, J. Chalain, D. Geoffroy, F. Laslaz, and G. Perrin, "The ALMA band-7 cartridge," *IEEE Transactions on Terahertz Science and Technology*, vol. 2, no. 1, pp. 29–39, 2012.

[35] Y. Sekimoto, Y. Iizuka, N. Satou, T. Ito, K. Kumagai, M. Kamikura, M. Naruse, and W. Shan, "Development of ALMA band 8 (385-500 GHz) cartridge," in *Proc. 19th Int. Symp. Space Terahertz Technol.*, pp. 253–257, 2008.

[36] A. Baryshev, R. Hesper, F. Mena, T. Klapwijk, T. Van Kempen, M. Hogerheijde, B. Jackson, J. Adema, G. Gerlofsma, M. Bekema, *et al.*, "The ALMA Band 9 receiver-Design, construction, characterization, and first light," *Astronomy & Astrophysics*, vol. 577, p. A129, 2015.

[37] S. Asayama, T. Takahashi, K. Kubo, T. Ito, M. Inata, T. Suzuki, T. Wada, T. Soga, C. Kamada, M. Karatsu, Y. Fujii, Y. Obuchi, S. Kawashima, H. Iwashita, and Y. Uzawa, "Development of ALMA Band 4 (125-163 GHz) receiver," *Publications of the Astronomical Society of Japan*, vol. 66, 06 2014.

[38] D. M. Pozar, *Microwave engineering*. John Wiley & Sons, 2009.

[39] J. Tucker, "Predicted conversion gain in superconductor-insulator-superconductor quasiparticle mixers," *Applied Physics Letters*, vol. 36, no. 6, pp. 477–479, 1980.

[40] T.-M. Shen, P. Richards, R. E. Harris, and F. Lloyd, "Conversion gain in mm-wave quasiparticle heterodyne mixers," *Applied Physics Letters*, vol. 36, no. 9, pp. 777–779, 1980.

[41] A. Smith, W. McGrath, P. Richards, H. Van Kempen, D. Prober, and P. Santhanam, "Negative resistance and conversion gain in SIS mixers," *Physica B+ C*, vol. 108, no. 1-3, pp. 1367–1368, 1981.

[42] R. N. Simons, *Coplanar waveguide circuits, components, and systems*, vol. 165. John Wiley & Sons, 2004.

[43] M. Maeda, "An analysis of gap in microstrip transmission lines," *IEEE Transactions on Microwave Theory and Techniques*, vol. 20, no. 6, pp. 390–396, 1972.

[44] R. E. Collin, *Foundations for microwave engineering*. John Wiley & Sons, 2007.

[45] H. A. Atwater, "MIcrostrip Reactive Circuit Elements," *IEEE Transactions on Microwave Theory and Techniques*, vol. 31, no. 6, pp. 488–491, 1983.

[46] R. Sorrentino and L. Roselli, "A new simple and accurate formula for microstrip radial stub," *IEEE Microwave and guided wave Letters*, vol. 2, no. 12, pp. 480–482, 1992.

[47] F. Giannini, R. Sorrentino, and J. Vrba, "Planar circuit analysis of microstrip radial stub (short paper)," *IEEE transactions on microwave theory and techniques*, vol. 32, no. 12, pp. 1652–1655, 1984.

[48] J. Kooi, M. Chan, B. Bumble, H. LeDuc, P. Schaffer, and T. Phillips, "230 and 492 GHz low noise sis waveguide receivers employing tuned Nb/AlO x/Nb tunnel junctions," *International journal of infrared and millimeter waves*, vol. 16, no. 12, pp. 2049–2068, 1995.

[49] J. W. Kooi, A. Kovács, M. C. Sumner, G. Chattopadhyay, R. Ceria, D. Miller, B. Bumble, H. G. LeDuc, J. A. Stern, and T. G. Phillips, "A 275–425-GHz tunerless waveguide receiver based on AlN-barrier SIS technology," *IEEE Transactions on Microwave Theory and Techniques*, vol. 55, no. 10, pp. 2086–2096, 2007.

[50] J. W. Kooi, C. K. Walker, H. LeDuc, P. Schaffer, T. Hunter, D. Benford, and T. Phillips, "A lownoise 665 GHz SIS quasi-particle waveguide receiver," *International journal of infrared and millimeter waves*, vol. 15, no. 3, pp. 477–492, 1994.

[51] J. Kraus and R. Marhefka, *Antennas for all applications*. McGraw-Hill series in electrical engineering, McGraw-Hill, 2002.

[52] W. Stutzman and G. Thiele, *Antenna Theory and Design*. Antenna Theory and Design, Wiley, 2012.

[53] J. Detlefsen and U. Siart, *Grundlagen der Hochfrequenztechnik*. Oldenbourg Verlag, 2010.

[54] M. J. M. van der Vorst, *Integrated lens antennas for submillimetre-wave applications*. Technische Universiteit Eindhoven, 1999.

[55] G. M. Rebeiz, "Millimeter-wave and terahertz integrated circuit antennas," *Proceedings of the IEEE*, vol. 80, no. 11, pp. 1748–1770, 1992.

[56] C. A. Balanis, "Antenna theory: A review," *Proceedings of the IEEE*, vol. 80, no. 1, pp. 7–23, 1992.

[57] V. P. Koshelets, S. V. Shitov, L. V. Filippenko, A. M. Baryshev, H. Golstein, T. de Graauw, W. Luinge, H. Schaeffer, and H. van de Stadt, "First implementation of a superconducting integrated receiver at 450 GHz," *Applied physics letters*, vol. 68, no. 9, pp. 1273–1275, 1996.

[58] V. P. Koshelets, S. V. Shitov, A. B. Ermakov, L. V. Filippenko, O. V. Koryukin, A. V. Khudchenko, M. Y. Torgashin, P. A. Yagoubov, R. W. Hoogeveen, and O. M. Pylypenko, "Superconducting integrated receiver for TELIS," *IEEE transactions on applied superconductivity*, vol. 15, no. 2, pp. 960–963, 2005.

[59] F. London and H. London, "The electromagnetic equations of the supraconductor," *Proc. R. Soc. Lond. A*, vol. 149, no. 866, pp. 71–88, 1935.

[60] J. Bardeen, L. N. Cooper, and J. R. Schrieffer, "Theory of superconductivity," *Physical Review*, vol. 108, no. 5, p. 1175, 1957.

[61] H. Fröhlich, "Theory of the superconducting state. I. The ground state at the absolute zero of temperature," *Physical Review*, vol. 79, no. 5, p. 845, 1950.

[62] M. Tinkham, *Introduction to superconductivity*. Courier Corporation, 2004.

[63] Y. Noat, V. Cherkez, C. Brun, T. Cren, C. Carbillet, F. Debontridder, K. Ilin, M. Siegel, A. Semenov, H.-W. Hübers, *et al.*, "Unconventional superconductivity in ultrathin superconducting NbN films studied by scanning tunneling spectroscopy," *Physical Review B*, vol. 88, no. 1, p. 014503, 2013.

[64] A. Abrikosov, "On the magnetic properties of a superconductor of the second kind," *Sov. Physics JETP*, vol. 5, p. 174, 1957.

[65] K. Ilin, D. Rall, M. Siegel, A. Engel, A. Schilling, A. Semenov, and H.-W. Huebers, "Influence of thickness, width and temperature on critical current density of Nb thin film structures," *Physica C: Superconductivity*, vol. 470, no. 19, pp. 953–956, 2010.

[66] C. Peroz and C. Villard, "Flux flow properties of niobium thin films in clean and dirty superconducting limits," *Physical Review B*, vol. 72, no. 1, p. 014515, 2005.

[67] V. Yurchenko, K. Ilin, J. Meckbach, M. Siegel, A. Qviller, Y. Galperin, and T. H. Johansen, "Thermo-magnetic stability of superconducting films controlled by nano-morphology," *Applied Physics Letters*, vol. 102, no. 25, p. 252601, 2013.

[68] C. Gorter, "Chapter I The Two Fluid Model for Superconductors and Helium II," in *Progress in Low Temperature Physics*, vol. 1, pp. 1–16, Elsevier, 1955.

[69] D. Mattis and J. Bardeen, "Theory of the anomalous skin effect in normal and superconducting metals," *Physical Review*, vol. 111, no. 2, p. 412, 1958.

[70] J. M. Pond, J. H. Claassen, and W. L. Carter, "Measurements and modeling of kinetic inductance microstrip delay lines," *IEEE transactions on microwave theory and techniques*, vol. 35, no. 12, pp. 1256–1262, 1987.

[71] B. D. Josephson, "Possible new effects in superconductive tunnelling," *Physics letters*, vol. 1, no. 7, pp. 251–253, 1962.

[72] P. W. Anderson and J. M. Rowell, "Probable observation of the Josephson superconducting tunneling effect," *Physical Review Letters*, vol. 10, no. 6, p. 230, 1963.

[73] D. Olaya, P. D. Dresselhaus, and S. P. Benz, "Niobium-silicide junction technology for superconducting digital electronics," *IEICE transactions on electronics*, vol. 93, no. 4, pp. 463–467, 2010.

[74] P.-G. De Gennes, *Superconductivity of metals and alloys*. CRC Press, 2018.

[75] A. Barone and G. Paterno, *Physics and applications of the Josephson effect*. Wiley, 1982.

[76] P. Gregers-Hansen and M. Levinsen, "Normal-state resistance as the determining parameter in the behavior of Dayem bridges with sinusoidal current-phase relations," *Physical Review Letters*, vol. 27, no. 13, p. 847, 1971.

[77] R. P. Feynman, R. B. Leighton, and M. Sands, "The feynman lectures on physics; vol. i," *American Journal of Physics*, vol. 33, no. 9, pp. 750–752, 1965.

[78] C. A. Hamilton, "Josephson voltage standards," *Review of scientific instruments*, vol. 71, no. 10, pp. 3611–3623, 2000.

[79] R. Gross, A. Marx, and F. Deppe, *Applied superconductivity: Josephson effect and superconducting electronics*. De Gruyter, 2015.

[80] V. Ambegaokar and A. Baratoff, "Tunneling between superconductors," *Physical Review Letters*, vol. 10, no. 11, p. 486, 1963.

[81] W. Buckel and R. Kleiner, *Superconductivity: fundamentals and applications*. John Wiley & Sons, 2008.

[82] S. T. Ruggiero, *Superconducting devices*. Elsevier, 2013.

[83] M. Octavio, M. Tinkham, G. Blonder, and T. Klapwijk, "Subharmonic energy-gap structure in superconducting constrictions," *Physical Review B*, vol. 27, no. 11, p. 6739, 1983.

[84] S. Rudner, M. Feldman, E. Kollberg, and T. Claeson, "Superconductor-insulator-superconductor mixing with arrays at millimeter-wave frequencies," *Journal of Applied Physics*, vol. 52, no. 10, pp. 6366–6376, 1981.

[85] P. Tien and J. Gordon, "Multiphoton process observed in the interaction of microwave fields with the tunneling between superconductor films," *Physical Review*, vol. 129, no. 2, p. 647, 1963.

[86] H. Ogawa, A. Mizuno, H. Hoko, H. Ishikawa, and Y. Fukui, "A 110 GHz SIS receiver for radio astronomy," *International Journal of Infrared and Millimeter Waves*, vol. 11, no. 6, pp. 717–726, 1990.

[87] A. Kerr, S.-K. Pan, M. Feldman, and A. Davidson, "Infinite available gain in a 115 GHz SIS mixer," *Physica B+ C*, vol. 108, no. 1-3, pp. 1369–1370, 1981.

[88] J. Zmuidzinas, J. W. Kooi, J. Kawamura, G. Chattopadhyay, B. Bumble, H. G. LeDuc, and J. A. Stern, "Development of SIS mixers for 1 THz," in *Advanced Technology MMW, Radio, and Terahertz Telescopes*, vol. 3357, pp. 53–63, International Society for Optics and Photonics, 1998.

[89] M. Westig, S. Selig, K. Jacobs, T. Klapwijk, and C. Honingh, "Improved Nb SIS devices for heterodyne mixers between 700 GHz and 1.3 THz with NbTiN transmission lines using a normal metal energy relaxation layer," *Journal of Applied Physics*, vol. 114, no. 12, p. 124504, 2013.

[90] M. Bin, M. Gaidis, J. Zmuidzinas, T. Phillips, and H. LeDuc, "Low-noise 1 THz niobium superconducting tunnel junction mixer with a normal metal tuning circuit," *Applied physics letters*, vol. 68, no. 12, pp. 1714–1716, 1996.

[91] M. J. Feldman, "Theoretical considerations for THz SIS mixers," *International journal of infrared and millimeter waves*, vol. 8, no. 10, pp. 1287–1292, 1987.

[92] G. Pance and M. J. Wengler, "Integrated tuning elements for millimeter and submillimeter SIS mixers," in *1992 IEEE MTT-S Microwave Symposium Digest*, pp. 337–340, IEEE, 1992.

[93] Y. Zhang and D. Winkler, "An integrated superconducting sub-mm wave receiver for linewidth measurements of Josephson flux-flow oscillators," *IEEE transactions on microwave theory and techniques*, vol. 42, no. 4, pp. 726–733, 1994.

[94] Z. Wang, Y. Uzawa, and A. Kawakami, "High current density NbN/AlN/NbN tunnel junctions for submillimeter wave SIS mixers," *IEEE transactions on applied superconductivity*, vol. 7, no. 2, pp. 2797–2800, 1997.

[95] P. Yadranjee Aghdam, H. Rashid, A. Pavolotsky, V. Desmaris, D. Meledin, and V. Belitsky, "On Specific Capacitance of SIS Junctions," in *26TH INTERNATIONAL SYMPOSIUM ON SPACE TERAHERTZ TECHNOLOGY*, 2015.

[96] J. Kawamura, D. Miller, J. Chen, J. Zmuidzinas, B. Bumble, H. G. LeDuc, and J. A. Stern, "Very high-current-density Nb/AlN/Nb tunnel junctions for low-noise submillimeter mixers," *Applied Physics Letters*, vol. 76, no. 15, pp. 2119–2121, 2000.

[97] A. Scott, "Distributed device applications of the superconducting tunnel junction," *Solid-State Electronics*, vol. 7, no. 2, pp. 137–146, 1964.

[98] T. Nagatsuma, K. Enpuku, F. Irie, and K. Yoshida, "Flux-flow type Josephson oscillator for millimeter and submillimeter wave region," *Journal of Applied Physics*, vol. 54, no. 6, pp. 3302–3309, 1983.

[99] J. J. Mazo and A. V. Ustinov, "The sine-Gordon equation in Josephson-Junction arrays," in *The sine-Gordon Model and its Applications*, pp. 155–175, Springer, 2014.

[100] J. C. Swihart, "Field solution for a thin-film superconducting strip transmission line," *Journal of Applied Physics*, vol. 32, no. 3, pp. 461–469, 1961.

[101] D. Coon and M. D. Fiske, "Josephson ac and step structure in the supercurrent tunneling characteristic," *Physical Review*, vol. 138, no. 3A, p. A744, 1965.

[102] V. P. Koshelets and J. Mygind, "Flux flow oscillators for superconducting integrated submm wave receivers," *Studies of High Temperature Superconductors*, pp. 213–246, 2001.

[103] T. Nagatsuma, K. Enpuku, K. Yoshida, and F. Irie, "Flux-flow-type Josephson oscillator for millimeter and submillimeter wave region. II. Modeling," *Journal of applied physics*, vol. 56, no. 11, pp. 3284–3293, 1984.

[104] V. Koshelets, P. Dmitriev, A. Sobolev, A. Pankratov, V. Khodos, V. Vaks, A. Baryshev, P. Wesselius, and J. Mygind, "Line width of Josephson flux flow oscillators," *Physica C: Superconductivity*, vol. 372, pp. 316–321, 2002.

[105] J. Mygind, V. Koshelets, M. R. Samuelsen, and A. Sobolev, "The submm wave Josephson Flux Flow Oscillator; linewidth measurements and simple theory," *IEEE transactions on applied superconductivity*, vol. 15, no. 2, pp. 968–971, 2005.

[106] V. Koshelets, S. Shitov, A. Shchukin, L. Filippenko, J. Mygind, and A. Ustinov, "Self-pumping effects and radiation linewidth of Josephson flux-flow oscillators," *Physical Review B*, vol. 56, no. 9, p. 5572, 1997.

[107] V. P. Koshelets, P. N. Dmitriev, A. B. Ermakov, A. S. Sobolev, A. M. Baryshev, P. R. Wesselius, and J. Mygind, "Radiation linewidth of flux-flow oscillators," *Superconductor Science and Technology*, vol. 14, no. 12, p. 1040, 2001.

[108] V. Koshelets, S. Shitov, L. Filippenko, P. Dmitriev, A. B. Ermakov, A. Sobolev, M. Y. Torgashin, A. Pankratov, V. Kurin, P. Yagoubov, *et al.*, "A superconducting phase-locked local oscillator for a submillimetre integrated receiver," *Superconductor Science and Technology*, vol. 17, no. 5, p. S127, 2004.

[109] V. P. Koshelets, P. N. Dmitriev, A. B. Ermakov, A. S. Sobolev, M. Y. Torgashin, V. V. Kurin, A. L. Pankratov, and J. Mygind, "Optimization of the phase-locked flux-flow oscillator for the submm integrated receiver," *IEEE transactions on applied superconductivity*, vol. 15, no. 2, pp. 964–967, 2005.

[110] K. Yoshida and F. Irie, "Frequency conversion in a long Josephson junction with a moving vortex array," *Applied Physics Letters*, vol. 27, no. 8, pp. 469–470, 1975.

[111] C. Tong, R. Blundell, B. Bumble, J. Stern, and H. LeDuc, "Distributed quasiparticle mixing in a non-linear transmission line at sub-millimeter wavelengths," *IEEE transactions on applied superconductivity*, vol. 7, no. 2, pp. 3597–3600, 1997.

[112] A. Benabdallah, J. Caputo, and A. C. Scott, "Exponentially tapered Josephson flux-flow oscillator," *Physical Review B*, vol. 54, no. 22, p. 16139, 1996.

[113] V. P. Koshelets, A. B. Ermakov, S. V. Shitov, P. Dmitriev, L. V. Filippenko, A. M. Baryshev, W. Luinge, J. Mygind, V. L. Vaks, and D. G. Pavel'ev, "Superfine resonant structure on IV-curves of long Josephson junction and its influence on flux flow oscillator linewidth," *IEEE transactions on applied superconductivity*, vol. 11, no. 1, pp. 1211–1214, 2001.

[114] T. Nagatsuma, K. Enpuku, K. Sueoka, K. Yoshida, and F. Irie, "Flux-flow-type Josephson oscillator for millimeter and submillimeter wave region. III. Oscillation stability," *Journal of applied physics*, vol. 58, no. 1, pp. 441–449, 1985.

[115] C. Kaiser, J. Meckbach, K. Ilin, J. Lisenfeld, R. Schäfer, A. Ustinov, and M. Siegel, "Aluminum hard mask technique for the fabrication of high quality submicron Nb/Al–AlOx/Nb josephson junctions," *Superconductor science and technology*, vol. 24, no. 3, p. 035005, 2010.

[116] C. Kaiser, *High Quality Nb Al-AlOx Nb Josephson Junctions: Technological Development and Macroscopic Quantum Experiments*, vol. 4. KIT Scientific Publishing, 2011.

[117] J. M. Meckbach, *Superconducting Multilayer Technology for Josephson Devices: Technology, Engineering, Physics, Applications*, vol. 13. KIT Scientific Publishing, 2013.

[118] S. K. Tolpygo, D. Amparo, D. T. Yohannes, M. Meckbach, and A. F. Kirichenko, “Process-Induced Variability of Nb/Al-AlOx/Nb Junctions in Superconductor Integrated Circuits and Protection Against It,” *IEEE Transactions on Applied Superconductivity*, vol. 19, no. 3, pp. 135–139, 2009.

[119] S. K. Tolpygo, D. Amparo, R. Hunt, J. Vivalda, and D. Yohannes, “Subgap Leakage in Nb/Al-AlOx/Nb Josephson Junctions and Run-to-Run Reproducibility: Effects of Oxidation Chamber and Film Stress,” *IEEE Transactions on Applied Superconductivity*, vol. 23, no. 3, pp. 1100305–1100305, 2013.

[120] S. K. Tolpygo and D. Amparo, “Fabrication-process-induced variations of Nb/Al/AlOx/Nb Josephson junctions in superconductor integrated circuits,” *Superconductor Science and Technology*, vol. 23, no. 3, p. 034024, 2010.

[121] R. Miller, W. Mallison, A. Kleinsasser, K. Delin, and E. Macedo, “Niobium trilayer Josephson tunnel junctions with ultrahigh critical current densities,” *Applied physics letters*, vol. 63, no. 10, pp. 1423–1425, 1993.

[122] H. Sugiyama, A. Fujimaki, and H. Hayakawa, “Characteristics of high critical current density Josephson junctions with Nb/AlO_x/Nb trilayers,” *IEEE Transactions on Applied Superconductivity*, vol. 5, no. 2, pp. 2739–2742, 1995.

[123] A. Kleinsasser, W. Mallison, and R. Miller, “Nb/AlN/Nb Josephson junctions with high critical current density,” *IEEE Transactions on Applied Superconductivity*, vol. 5, no. 2, pp. 2318–2321, 1995.

[124] H. Kroger, L. Smith, and D. Jillie, “Selective niobium anodization process for fabricating Josephson tunnel junctions,” *Applied Physics Letters*, vol. 39, no. 3, pp. 280–282, 1981.

[125] S. Anders, M. Schmelz, L. Fritzsch, R. Stolz, V. Zakosarenko, T. Schönau, and H. Meyer, “Sub-micrometer-sized, cross-type Nb–AlOx–Nb tunnel junctions with low parasitic capacitance,” *Superconductor Science and Technology*, vol. 22, no. 6, p. 064012, 2009.

[126] J. Yong, T. Lemberger, L. Benfatto, K. Ilin, and M. Siegel, “Robustness of the Berezinskii-Kosterlitz-Thouless transition in ultrathin NbN films near the superconductor-insulator transition,” *Physical Review B*, vol. 87, no. 18, p. 184505, 2013.

[127] Z. Wang, A. Kawakami, Y. Uzawa, and B. Komiyama, "Superconducting properties and crystal structures of single-crystal niobium nitride thin films deposited at ambient substrate temperature," *Journal of applied physics*, vol. 79, no. 10, pp. 7837–7842, 1996.

[128] Z. Wang, H. Terai, A. Kawakami, and Y. Uzawa, "Interface and tunneling barrier heights of NbN/AlN/NbN tunnel junctions," *Applied physics letters*, vol. 75, no. 5, pp. 701–703, 1999.

[129] K. Ilin, R. Schneider, D. Gerthsen, A. Engel, H. Bartolf, A. Schilling, A. Semenov, H. Huebers, B. Freitag, and M. Siegel, "Ultra-thin NbN films on Si: crystalline and superconducting properties," in *Journal of Physics: Conference Series*, vol. 97, p. 012045, IOP Publishing, 2008.

[130] H. Akaike, T. Funai, S. Sakamoto, and A. Fujimaki, "Fabrication of NbN/Al-AlN x/NbN tunnel junctions on several kinds of substrates," in *2013 IEEE 14th International Superconductive Electronics Conference (ISEC)*, pp. 1–3, IEEE, 2013.

[131] A. Kawakami, Z. Wang, and S. Miki, "Fabrication and characterization of epitaxial NbN/MgO/NbN Josephson tunnel junctions," *Journal of Applied Physics*, vol. 90, no. 9, pp. 4796–4799, 2001.

[132] Z. Wang, A. Kawakami, Y. Uzawa, and B. Komiyama, "High critical current density NbN/AlN/NbN tunnel junctions fabricated on ambient temperature MgO substrates," *Applied physics letters*, vol. 64, no. 15, pp. 2034–2036, 1994.

[133] Z. Wang, A. Kawakami, and Y. Uzawa, "NbN/AlN/NbN tunnel junctions with high current density up to 54 kA/cm 2," *Applied physics letters*, vol. 70, no. 1, pp. 114–116, 1997.

[134] K. Makise, H. Terai, and Y. Uzawa, "NbN/AlN/NbN/TiN tunnel junctions on Si (100) substrate for superconducting devices," *IEEE Transactions on Applied Superconductivity*, vol. 26, no. 3, pp. 1–3, 2016.

[135] J. Villegier, L. Vieux-Rochaz, M. Goniche, P. Renard, and M. Vabre, "NbN tunnel junctions," *IEEE Transactions on Magnetics*, vol. 21, no. 2, pp. 498–504, 1985.

[136] *GoldExi.* [online]: `http://www.oocities.org/goldexi/`. 2019-01-12.

[137] T. Gaber, E. Goldobin, A. Sterck, R. Kleiner, D. Koelle, M. Siegel, and M. Neuhaus, "Nonideal artificial phase discontinuity in long Josephson 0- kappa junctions," *Physical Review B*, vol. 72, no. 5, p. 054522, 2005.

[138] E. Goldobin, A. Sterck, T. Gaber, D. Koelle, and R. Kleiner, "Dynamics of semifluxons in Nb long Josephson 0- pi junctions," *Physical review letters*, vol. 92, no. 5, p. 057005, 2004.

[139] T. Shiino, S. Shiba, N. Sakai, T. Yamakura, L. Jiang, Y. Uzawa, H. Maezawa, and S. Yamamoto, "Improvement of the critical temperature of superconducting NbTiN and NbN thin films using the AlN buffer layer," *Superconductor Science and Technology*, vol. 23, no. 4, p. 045004, 2010.

[140] *CST-computer simulation technology gmbh.* [online]: `http://www.CST.com`, 2018-10-19.

[141] A. Kawakami, S. Miki, Y. Uzawa, and Z. Wang, "Estimation of surface resistance for epitaxial NbN films in the frequency range of 0.1-1.1 THz," *IEEE transactions on applied superconductivity*, vol. 13, no. 2, pp. 1147–1150, 2003.

[142] S. Shitov, A. Ustinov, N. Iosad, and H. Kohlstedt, "On-chip radiation detection from stacked Josephson flux-flow oscillators," *Journal of applied physics*, vol. 80, no. 12, pp. 7134–7137, 1996.

[143] S. Shitov, V. P. Koshelets, A. Baryshev, I. Lapitskaya, L. Filippenko, T. de Graauw, H. Schaeffer, H. van de Stadt, and W. Luinge, "A Superconducting Planar Integrated Receiver for the Frequency Range 430-480 GHz," in *Proceedings of 6th Space TeraHertz Technology Conference, Pasadena*, 1995.

[144] J. Zmuidzinas, H. G. LeDuc, J. A. Stern, and S. R. Cypher, "Two-junction tuning circuits for submillimeter SIS mixers," *IEEE transactions on microwave theory and techniques*, vol. 42, no. 4, pp. 698–706, 1994.

[145] B. S. Karasik, A. V. Sergeev, and D. E. Prober, "Nanobolometers for THz photon detection," *IEEE Transactions on Terahertz Science and Technology*, vol. 1, no. 1, pp. 97–111, 2011.

[146] A. Scheuring, *Ultrabreitbandige Strahlungseinkopplung in THz-Detektoren*, vol. 11. KIT Scientific Publishing, 2014.

[147] K. Ilin, D. Henrich, Y. Luck, Y. Liang, M. Siegel, and D. Y. Vodolazov, "Critical current of Nb, NbN, and TaN thin-film bridges with and without geometrical nonuniformities in a magnetic field," *Physical Review B*, vol. 89, no. 18, p. 184511, 2014.

[148] I. Charaev, A. Semenov, S. Doerner, G. Gomard, K. Ilin, and M. Siegel, "Current dependence of the hot-spot response spectrum of superconducting single-photon detectors with different layouts," *Superconductor Science and Technology*, vol. 30, no. 2, p. 025016, 2016.

[149] J. R. Clem and K. K. Berggren, "Geometry-dependent critical currents in superconducting nanocircuits," *Physical Review B*, vol. 84, no. 17, p. 174510, 2011.

[150] A. Scheuring, P. Thoma, J. Day, K. Il'in, J. Hanisch, B. Holzapfel, and M. Siegel, "Thin Pr–Ba–Cu–O film antenna-coupled THz bolometers for room temperature operation," *IEEE Transactions on Terahertz Science and Technology*, vol. 3, no. 1, pp. 103–109, 2013.

[151] D. M. Slocum, T. M. Goyette, E. J. Slingerland, R. H. Giles, and W. E. Nixon, "Terahertz atmospheric attenuation and continuum effects," in *Terahertz Physics, Devices, and Systems VII: Advanced Applications in Industry and Defense*, vol. 8716, p. 871607, International Society for Optics and Photonics, 2013.

Publications within this thesis

[AWG$^+$17] M. Arndt, S. Wuensch, C. Groetsch, M. Merker, G. Zieger, K. Peiselt, S. Anders, H.-G. Meyer, and M. Siegel. Optimization of the microwave properties of the kinetic-inductance bolometer (kibo). *IEEE Transactions on Applied Superconductivity*, 2017.

[BSU$^+$18] S. Bagheri, N. Strohfeldt, M. Ubl, A. Berrier, M. Merker, G. Richter, M. Siegel, and H. Giessen. Niobium as alternative material for refractory and active plasmonics. *ACS Photonics*, 2018.

[KSS$^+$15] A. A. Kuzmin, A. D. Semenov, S. V. Shitov, M. Merker, S. H. Wuensch, A. V. Ustinov, and M. Siegel. Superconducting noise bolometer with microwave bias and readout for array applications. *IEEE Transactions on Applied Superconductivity*, 25(3):1–4, 2015.

[MBV$^+$17] M. Merker, C. Bohn, M. Völlinger, K. Ilin, and M. Siegel. Nbn/aln/nbn josephson junctions on sapphire for sis receiver applications. *IEEE Transactions on Applied Superconductivity*, 27(4):1–5, 2017.

[MMB$^+$13] J.M. Meckbach, M. Merker, S.J. Buehler, K. Ilin, B. Neumeier, U. Kienzle, E. Goldobin, R. Kleiner, D. Koelle, and M. Siegel. Sub-μm josephson junctions for superconducting quantum devices. *IEEE Transactions on Applied Superconductivity*, 23(3):1100504–1100504, 2013.

[MMS$^+$18] R. Menditto, M. Merker, M. Siegel, D. Koelle, R. Kleiner, and E. Goldobin. Evidence of macroscopic quantum tunneling from both wells in a φ josephson junction. *Physical Review B*, 98(2):024509, 2018.

[SAK$^+$15] S. V. Shitov, N. N. Abramov, A. A. Kuzmin, M. Merker, M. Arndt, S. H. Wuensch, K. Ilin, E. Erhan, A. V. Ustinov, and M. Siegel. Wide-range bolometer with rf readout tes. *IEEE Transactions on Applied Superconductivity*, 25(3):1–4, 2015.

[SKM+17] S. V. Shitov, A. A. Kuzmin, M. Merker, V. I. Chichkov, A. V. Merenkov, A. B. Ermakov, A. V. Ustinov, and M. Siegel. Progress in development of the superconducting bolometer with microwave bias and readout. *IEEE Transactions on Applied Superconductivity*, 2017.

[VIMS15] D. Y. Vodolazov, K. Ilin, M. Merker, and M. Siegel. Defect-controlled vortex generation in current-carrying narrow superconducting strips. *Superconductor Science and Technology*, 29(2):025002, 2015.

[WGMS15] S. Wuensch, C. Groetsch, M. Merker, and M. Siegel. Optimized lekid structures with low crosstalk for large detector arrays. *IEEE Transactions on Applied Superconductivity*, 25(3):1–5, 2015.

Supervised student thesis

[Boh14] BOHN, Christian: *Herstellung von NbN/AlN/NbN Trilayern bei Raumtemperatur und Optimierung der supraleitenden Tunneleigenschaften*, Karlsruhe Institute of Technology (KIT), Insitute of micro- and nanoelectronic Systems (IMS), Bachelors thesis, 2014

[Boh17] BOHN, Christian: *Untersuchung der Ankopplung von Flux-Flow Oszillatoren an SIS-Mischer basierend auf NbN/AlN/NbN Trilayern*, Karlsruhe Institute of Technology (KIT), Insitute of micro- and nanoelectronic Systems (IMS), Masters thesis, 2017

[Bre14] BRENNER, Sina: *Entwicklung und Aufbau eines kompakten Infrarotsystems zur Detektion von Gefahrstoffen*, Karlsruhe Institute of Technology (KIT), Insitute of micro- and nanoelectronic Systems (IMS), Masters thesis, 2014

[Bus15] BUSCH, Tobias: *Implementierung von Inline-Messungen mittels Reflektometrie in einem MEMS-Prozess zur Sicherstellung der Fertigungsstabilität*, Karlsruhe Institute of Technology (KIT), Insitute of micro- and nanoelectronic Systems (IMS), Masters thesis, 2015

[Dis13] DISCHKE, Christian: *Entwicklung einer automatisierten Steuerung für ein THz-Sensorsystem*, Karlsruhe Institute of Technology (KIT), Insitute of micro- and nanoelectronic Systems (IMS), Diploma thesis, 2013

[Fis15] FISCHER, Karen: *Ermittlung der Hochfrequenzeingenschaften von NbN-Dünnschichtfilmen am IMS*, Karlsruhe Institute of Technology (KIT), Insitute of micro- and nanoelectronic Systems (IMS), Bachelors thesis, 2015

[Kar13] KARCHER, Christian: *Simulation und Fabrikation von Lokaloszillatoren basierend auf langen Josephsonkontakten*, Karlsruhe Institute of Technology (KIT), Insitute of micro- and nanoelectronic Systems (IMS), Masters thesis, 2013

[Mar16] MARHOLZ, Malte: *Entwurf und Simulation eines Antennendesigns für einen SIS-Mischer im THz-Bereich*, Karlsruhe Institute of Technology (KIT), Insitute of micro- and nanoelectronic Systems (IMS), Bachelors thesis, 2016

[Mat15] MATURA, Markus: *AlN/NbN/AlN multi-layer for detector applications*, Karlsruhe Institute of Technology (KIT), Insitute of micro- and nanoelectronic Systems (IMS), Studienarbeit, 2015

[Non16] NONNENMACHER, Friedemann: *Untersuchung der Herstellungsprozesses von NbN/AlN/NbN Josephson Kontakten auf MgO*, Karlsruhe Institute of Technology (KIT), Insitute of micro- and nanoelectronic Systems (IMS), Bachelors thesis, 2016

[Rei17] REISER, Lisa: *Untersuchung eines Koplanardesigns zur Realisierung eines integrierten Receivers für Frequenzen im THz Bereich*, Karlsruhe Institute of Technology (KIT), Insitute of micro- and nanoelectronic Systems (IMS), Masters thesis, 2017

[Sch18] SCHMID, Simon: *Entwicklung von hochsensitiven NbN Terahertz Bolometern durch thermische Entkopplung von Mikrobrücken*, Karlsruhe Institute of Technology (KIT), Insitute of micro- and nanoelectronic Systems (IMS), Masters thesis, 2018

[Völ16] VÖLLINGER, Marvin: *Realisierung von qualitativ hochwertigen NbN/AlN/NbN Josephson-Kontakten auf Magnesiumoxid*, Karlsruhe Institute of Technology (KIT), Insitute of micro- and nanoelectronic Systems (IMS), Bachelors thesis, 2016

International Conferences

- J. M. Meckbach, M. Merker, S. Bühler, K. Ilin, N. Benjamin, U. Kienzle, E. Goldobin, R. Kleiner, D. Kölle, and M. Siegel, "Sub-μ-m josephson junctions for superconducting quantum devices." In *Applied Superconductivity Conference, Portland, USA*, 2012.
- J. M. Meckbach, S. Bühler, M. Merker, K. Ilin, K. Buckenmaier, T. Gaber, U. Kienzle, B. Neumeier, E. Goldobin, R. Kleiner, and D. Kölle, "Versatile multi-layer josephson junction process for vortex molecules." In *DPG Frühjahrstagung, Berlin, Germany*, 2012.
- S. Bühler, J. M. Meckbach, S. Wünsch, M. Merker, K. Ilin, B. Neumeier, E. Goldobin, R. Kleiner, D. Kölle, and M. Siegel, "Creation of artificial 1d vortex-crystals using current injectors." In *Tagung Kryoelektronische Bauelemente, Freudenstadt, Germany*, 2012.
- M. Merker, J. M. Meckbach, K. Ilin, and M. Siegel, "Fabrication process for planarization and miniaturization of josephson junctions." In *Tagung Kryoelektronische Bauelemente, Freudenstadt, Germany*, 2012.
- J. M. Meckbach, A. Häffelin, S. Bühler, M. Merker, K. Ilin, and M. Siegel, "Niobium nitride technology for josephson junction devices." In *DPG Frühjahrstagung, Regensburg, Germany*, 2013.
- M. Merker, J. M. Meckbach, S. Bühler, K. Ilin, and M. Siegel, "Self-planarized process for the fabrication of josephson junction devices." In *DPG Frühjahrstagung, Regensburg, Germany*, 2013.
- J. M. Meckbach, M. Merker, S. Bühler, K. Ilin, and M. Siegel, "Multi-layer technology for nbn/aln/nbn josephson junctions." In *European Conference on applied superconductivity, Genova, Italy*, 2013.
- M. Merker, J. M. Meckbach, S. Bühler, M. Rudolph, K. Ilin, D. Kölle, R. Kleiner, and M. Siegel, "Planar multilayer process for josephson junction devices." In *International Superconductive Electronics Conference, Cambridge, USA*, 2013.

- M. Merker, J. M. Meckbach, S. Bühler, K. Ilin, and M. Siegel, "All-nbn/aln/nbn josephson junctions on silicon substrates." In *Tagung Kryoelektronische Bauelemente, Bad Herrenalb, Germany*, 2013.
- M. Merker, C. Bohn, K. Ilin, and M. Siegel, "Room temperature deposited nbn/aln/nbn trilayers on different substrates for josephson junctions." In *Tagung Kryoelektronische Bauelemente, Berlin, Germany*, 2014.
- E. Schmidt, M. Merker, K. Ilin, and M. Siegel, "Superconducting nbn single-photon detectors on gaas with an aln buffer layer." In *DPG Frühjahrstagung, Berlin, Germany*, 2015.
- M. Merker, C. Bohn, M. Marholz, A. Schmid, K. Ilin, and M. Siegel, "Nbn/aln/nbn josephson junction fabrication on oxidized silicon substrates." In *Tagung Kryoelektronische Bauelemente, Warberg, Germany*, 2015.
- A. Kuzmin, S. Shitov, M. Merker, M. Arndt, S. Wünsch, K. Ilin, A. Ustinov, and M. Siegel, "Scalable array of superconducting terahertz nb nanobolometers with microwave bias and readout." In *European Conference on applied superconductivity, Lyon, France*, 2015.
- M. Merker, C. Bohn, M. Völlinger, K. Ilin, and M. Siegel, "Nbn/aln/nbn josephson junctions on sapphire for sis receiver applications." In *Applied Superconductivity Conference, Denver, USA*, 2016.
- M. Merker, C. Bohn, M. Völlinger, K. Ilin, and M. Siegel, "Development of nbn/aln/nbn josephson junctions on sapphire substrates." In *Tagung Kryoelektronische Bauelemente, Freyburg, Germany*, 2016.
- M. Merker, C. Bohn, M. Völlinger, K. Ilin, and M. Siegel, "Nbn/aln/nbn josephson junctions on different substrates." In *DPG Frühjahrstagung, Regensburg, Germany*, 2016.
- M. Merker, A. Schmid, L. Reiser, M. Völlinger, K. Ilin, and M. Siegel, "Nbn/aln/nbn technology for sis recievers." In *International Superconductive Electronics Conference, Sorrento, Italy*, 2017.
- M. Merker, A. Schmid, K. Ilin, and M. Siegel, "Integration of a flux-flow oscillator with an sis mixer based on nbn/aln/nbn technology." In *Tagung Kryoelektronische Bauelemente, Heidelberg, Germany*, 2018.

- M. Merker, A. Schmid, C. Bohn, K. Ilin, and M. Siegel, "Nbn/aln/nbn technology on sapphire substrates for sis based thz receivers." In *29th IEEE International Symposium on Space THz Technology, Pasadena, California, USA*, 2018.

Karlsruher Schriftenreihe zur Supraleitung

Karlsruher Institut für Technologie (KIT) | ISSN 1869-1765

Herausgeber: Prof. Dr.-Ing. M. Noe, Prof. Dr. rer. nat. M. Siegel

Band 001 **Christian Schacherer**
Theoretische und experimentelle Untersuchungen zur Entwicklung supraleitender resistiver Strombegrenzer. 2009
ISBN 978-3-86644-412-6

Band 002 **Alexander Winkler**
Transient behaviour of ITER poloidal field coils. 2011
ISBN 978-3-86644-595-6

Band 003 **André Berger**
Entwicklung supraleitender, strombegrenzender Transformatoren. 2011
ISBN 978-3-86644-637-3

Band 004 **Christoph Kaiser**
High quality Nb/Al-AlOx/Nb Josephson junctions. Technological development and macroscopic quantum experiments. 2011
ISBN 978-3-86644-651-9

Band 005 **Gerd Hammer**
Untersuchung der Eigenschaften von planaren Mikrowellen-resonatoren für Kinetic-Inductance Detektoren bei 4,2 K. 2011
ISBN 978-3-86644-715-8

Band 006 **Olaf Mäder**
Simulationen und Experimente zum Stabilitätsverhalten von HTSL-Bandleitern. 2012
ISBN 978-3-86644-868-1

Die Bände sind unter www.ksp.kit.edu als PDF frei verfügbar oder als Druckausgabe bestellbar.

Band 007 **Christian Barth**
High Temperature Superconductor Cable Concepts for Fusion Magnets. 2013
ISBN 978-3-7315-0065-0

Band 008 **Axel Stockhausen**
Optimization of Hot-Electron Bolometers for THz Radiation. 2013
ISBN 978-3-7315-0066-7

Band 009 **Petra Thoma**
Ultra-fast $YBa_2Cu_3O_{7-x}$ direct detectors for the THz frequency range. 2013
ISBN 978-3-7315-0070-4

Band 010 **Dagmar Henrich**
Influence of Material and Geometry on the Performance of Superconducting Nanowire Single-Photon Detectors. 2013
ISBN 978-3-7315-0092-6

Band 011 **Alexander Scheuring**
Ultrabreitbandige Strahlungseinkopplung in THz-Detektoren. 2013
ISBN 978-3-7315-0102-2

Band 012 **Markus Rösch**
Development of lumped element kinetic inductance detectors for mm-wave astronomy at the IRAM 30 m telescope. 2013
ISBN 978-3-7315-0110-7

Band 013 **Johannes Maximilian Meckbach**
Superconducting Multilayer Technology for Josephson Devices. 2013
ISBN 978-3-7315-0122-0

Band 014 **Enrico Rizzo**
Simulations for the optimization of High Temperature Superconductor current leads for nuclear fusion applications. 2014
ISBN 978-3-7315-0132-9

Band 015 **Philipp Krüger**
Optimisation of hysteretic losses in high-temperature superconducting wires. 2014
ISBN 978-3-7315-0185-5

Band 016 **Matthias Hofherr**
Real-time imaging systems for superconducting nanowire single-photon detector arrays. 2014
ISBN 978-3-7315-0229-6

Band 017 **Oliver Näckel**
Development of an Air Coil Superconducting Fault Current Limiter. 2016
ISBN 978-3-7315-0526-6

Band 018 **Christoph M. Bayer**
Characterization of High Temperature Superconductor Cables for Magnet Toroidal Field Coils of the DEMO Fusion Power Plant. 2017
ISBN 978-3-7315-0605-8

Band 019 **Shengnan Zou**
Magnetization of High Temperature Superconducting Trapped-Field Magnets. 2017
ISBN 978-3-7315-0715-4

Band 020 **Ilya Charaev**
Improving the Spectral Bandwidth of Superconducting Nanowire Single-Photon Detectors (SNSPDs). 2018
ISBN 978-3-7315-0745-1

Band 021 **Juliane Raasch**
Electrical-field sensitive $YBa_2Cu_3O_{7-x}$ detectors for real-time monitoring of picosecond THz pulses. 2019
ISBN 978-3-7315-0786-4

Band 022 **Yingzhen Liu**
Design of a superconducting DC wind generator. 2019
ISBN 978-3-7315-0796-3

Band 023 **Sebastian Hellmann**
Research and Technology Development on Superconducting Current Limiting Transformers. 2019
ISBN 978-3-7315-0804-5

Band 024 **Simon J. Otten**
Characterisation of REBCO Roebel cables. 2019
ISBN 978-3-7315-0904-2

Band 025 **Julia Brandel**
Supraleitende Einzelphotonenzähler: Optimierung der Zeitauflösung und Anwendungsbeispiele aus der Spektroskopie. 2019
ISBN 978-3-7315-0917-2

Band 026 **Dustin Kottonau, Eugen Shabagin, Wescley T. B. de Sousa, Jörn Geisbüsch, Mathias Noe, Hanno Stagge, Simon Fechner, Hannes Woiton, Thomas Küsters**
Bewertung des Einsatzes supraleitender 380-kV-Kabel. 2019
ISBN 978-3-7315-0927-1

Band 027 **Steffen Dörner**
Multifrequenzausleseverfahren von supraleitenden Einzelphotonen-Detektoren. 2019
ISBN 978-3-7315-0961-5

Band 028 **Michael Merker**
Superconducting integrated THz receiver. 2019
ISBN 978-3-7315-0970-7